ELECTRIC PLASMAS:
THEIR NATURE AND USES

ELECTRIC PLASMAS:
THEIR NATURE AND USES

A. von Engel

Department of Engineering Science, University of Oxford
Honorary Fellow of Keble College, Oxford

Taylor & Francis Ltd
London and New York
1983

First published 1983 by Taylor & Francis Ltd,
4 John Street, London WC1N 2ET

Typeset by Gibbons Barford Print, Wednesfield Road,
Wolverhampton WV10 0JA
Printed in Great Britain by Taylor & Francis (Printers) Ltd,
Rankine Road, Basingstoke, Hampshire RG24 0PR

Distributed in the USA and Canada by
International Publications Service Taylor & Francis Inc.
114 East 32nd Street, New York, NY 10016

British Library Cataloguing in Publication Data

Von Engel, A.
   Electric plasmas.
   1. Plasmas (Ionized) gases
   I. Title
   530.4′4   QC718

   ISBN 0-85066-147-1

CONTENTS

Contents

PREFACE

Books on electric conduction in gases, ionization phenomena, gas discharges, gaseous electronics and plasma physics abound. What then induced me to write yet another text? There were two main motives for action: for some time the second edition (1965) of my book *Ionized Gases* has no longer been available, in spite of translations into Japanese, Russian and Yugoslavian; languages which neither I nor most students master. I also felt that there was a need for a text in which not only theory but also applied sciences are comprehensively treated. The Oxford University M.Sc. course in "Science and Applications of Electric Plasmas" in which I am engaged has facilitated the selection of problems and the allotment of space. I have attempted to present the subject in the form of extended essays in which elementary arguments are included for reasons best known only to those who actually teach it. I have deviated from the present fashion to collect and edit contributions of experts wallowing in their special fields, although this deprives the reader of accumulating scientific gems and possibly priceless revelations of future trends. On the other hand a 'single author' book provides the reader with a certain uniformity of approach, however biased and incomplete. Moreover, this book gives me the chance to tell a wider circle than that of my colleagues and pupils, various thoughts that I would not be able to publish in learned journals. I am eagerly awaiting letters from those who spot some of the novelties. However, no rewards can be claimed, neither for passing that exercise (modestly described as a treasure hunt) nor for detecting mistakes and omissions.

No one system of units is used consistently throughout. M.K.S. units are more common as much of the research has been, and still is done using those units, and is therefore easier to appreciate in that form. A Table of Units (p. xi) is included for reference, if required. Familiarity with spectroscopic nomenclature is assumed.

In teaching, politics and writing the common maxim holds that to know what to conceal is more important than to know what to disclose and thus, with much regret, I have had to exclude several items from the text. The axed subjects are: coherent and incoherent light sources, i.e., lasers and illumination; transient, pulsed and pulsing discharges; striations, stationary as well as moving; certain electro-mechanical effects; ion beam sources; and spectroscopic diagnostics. Though excluded from the text, references are given at the end of the last chapter. To keep the book to a reasonable length I have also excluded examples and their solutions.

Preface

To mix teaching, research and mini-administration with writing is spending time in a singularly inefficient way—or so I am told by those whose goddess is materialism. Few realize that in order to contribute to problems in physics ample time must be set aside to ponder about them so that the solution is to one's satisfaction. Looking back over the last years, I have enjoyed writing this text and I hope that this will percolate to the reader.

Fortunately I worked for and with a publisher who did not fix a precise delivery date for the manuscript. I am grateful for his advice and help and I owe much gratitude to Bob Noakes for numerous suggestions. The late Professor D. W. Holder, F.R.S. provided the facilities necessary to continue my research. My thanks go to my colleagues and research students for their tolerance and understanding. Raoul Franklin was a helpful pathfinder in the dense forest (or jungle) of publications, capable of providing instant clarification to countless scientific and non-academic questions.

The Warden and the Governing Body of Keble College, Oxford have continued to give me social and other facilities which I value highly and for which I express my thanks. My close relations with Culham Laboratory and some financial and material support is mainly due to Dr. R. S. Pease, F.R.S. whom I thank for his unfailing friendship. I am most grateful to my wife Ilse for painstaking secretarial help and to Dr. Peter Edgley who has cast a critical eye on the manuscript and has suggested various improvements and corrections. My thanks go to Mr. J. H. C. Maple, Culham Laboratory for advice and corrections for chapter 9, and to Dr. J. Cheney, the publisher's Scientific Editor, for his help, patience and counsel. I have benefited from being associated with Professors H. Motz and L. C. Woods, Dr. J. E. Allen, the initiator of the Plasma Course, and Professor K. G. Emeleus of Belfast University. I am indebted to Professor C. P. Wroth, Head of the Engineering Science Department, Col. R. H. Parsons, Administrator and Mrs. Esther Rose, Librarian, for rendering facilities and help which made my life and work a gentle burden. The illustrations have prospered in Mrs. J. Takacs' skilful hands. Finally I wish to remember Billy, my late feline companion for some 17 years, who gracefully acted as a live paperweight.

A.v.E.

PERIODIC TABLE OF THE ELEMENTS

1	2	3	4	5	6	7	8	9	10	11	12	13	14	15	16	17	18
1 H 1·0																	2 He 4·0
3 Li 6·9	4 Be 9·0											5 B 10·8	6 C 12·0	7 N 14·0	8 O 16·0	9 F 19·0	10 Ne 20·2
11 Na 23·0	12 Mg 24·3											13 Al 27·0	14 Si 28·1	15 P 31·0	16 S 32·1	17 Cl 35·5	18 Ar 39·9
19 K 39·1	20 Ca 40·1	21 Sc 45·0	22 Ti 47·9	23 V 50·9	24 Cr 52·0	25 Mn 54·9	26 Fe 55·8	27 Co 58·9	28 Ni 58·7	29 Cu 63·5	30 Zn 65·4	31 Ga 69·7	32 Ge 72·6	33 As 74·9	34 Se 79·0	35 Br 79·9	36 Kr 83·8
37 Rb 85·5	38 Sr 87·6	39 Y 88·9	40 Zr 91·2	41 Nb 92·9	42 Mo 95·9	43 Tc (99)	44 Ru 101·1	45 Rh 102·9	46 Pd 106·4	47 Ag 107·9	48 Cd 112·4	49 In 114·8	50 Sn 118·7	51 Sb 121·7	52 Te 127·6	53 I 126·9	54 Xe 131·3
55 Cs 132·9	56 Ba 137·3	57–71	72 Hf 178·5	73 Ta 180·9	74 W 183·8	75 Re 186·2	76 Os 190·2	77 Ir 192·2	78 Pt 195·1	79 Au 197·0	80 Hg 200·6	81 Tl 204·4	82 Pb 207·2	83 Bi 209·0	84 Po (210)	85 At (210)	86 Rn (222)
87 Fr (223)	88 Ra (227)	89–103	104 Rf	105 Ha													

Lanthanide series

57 La 138·9	58 Ce 140·1	59 Pr 140·9	60 Nd 144·2	61 Pm (145)	62 Sm 150·3	63 Eu 152·0	64 Gd 157·2	65 Tb 158·9	66 Dv 162·5	67 Ho 164·9	68 Er 167·3	69 Tm 168·9	70 Yb 173·0	71 Lu 175·0

Actinide series

89 Ac (227)	90 Th 232·0	91 Pa (231)	92 U 238·0	93 Np (237)	94 Pu (242)	95 Am (243)	96 Cm (245)	97 Bk (249)	98 Cf (249)	99 Es (254)	100 Fm (252)	101 Md (256)	102 No (254)	103 Lr

TABLE OF UNITS

Unit	Abbreviation	Equivalent
Ångström	Å	$1\ \text{Å} = 10^{-10}\ \text{m} = 0.1\ \text{nm}$
Atomic mass unit	a.m.u.	$1\ \text{a.m.u.} = 1\ \text{dalton} \simeq 1.6 \times 10^{-24}\ \text{g}$
Coulomb	C	$1\ \text{C} = 1\ \text{A s}$
Electron volt	eV	$1\ \text{eV} \simeq 1.6 \times 10^{-19}\ \text{J}$
Erg	erg	$1\ \text{erg} = 10^{-7}\ \text{J}$
Farad	F	$1\ \text{F} = 1\ \text{C V}^{-1}$
Gauss	G	$1\ \text{G} \triangleq 10^{-4}\ \text{T}$
Hertz	Hz	$1\ \text{Hz} = 1\ \text{cycle s}^{-1}$
Joule	J	$1\ \text{J} = 1\ \text{N m}$
Millimetre of mercury	mm Hg	$1\ \text{mm Hg} \simeq 1\ \text{Torr} \simeq 133\ \text{Pa}$
Nanometre	nm	$1\ \text{nm} = 10^{-9}\ \text{m} = 10\ \text{Å}$
Ohm	Ω	$1\ \Omega = 1\ \text{V A}^{-1}$
Pascal	Pa	$1\ \text{Pa} = 1\ \text{N m}^{-2}$
Poise	P	$1\ \text{P} = 0.1\ \text{Pa s}$
Standard atmosphere	atm	$1\ \text{atm} \simeq 101\ \text{kPa}$
Tesla	T	$1\ \text{T} = 1\ \text{V s m}^{-2}$
Torr	Torr	$1\ \text{Torr} \simeq 1\ \text{mm Hg} \simeq 133\ \text{Pa}$
Volt	V	$1\ \text{V} = 1\ \text{J A}^{-1}\ \text{s}^{-1}$
Watt	W	$1\ \text{W} = 1\ \text{J s}^{-1}$
Weber	Wb	$1\ \text{Wb} = 1\ \text{V s}^{-1}$

CHAPTER 1

introduction

The rapid advances in physics during this century have introduced a large number of new concepts. This fast expansion of our knowledge has been accompanied by a gradual disappearance of earlier well-established frontiers between the sciences. The boundaries between physics and chemistry, chemistry and engineering, the life sciences and the physical sciences have been torn down. This in turn has been accompanied by a division of the 'main' subjects into smaller and smaller sections. Today we talk about atomic physics, molecular physics, solid-state physics, nuclear physics and chemical physics to mention only a few, plasma physics being one of the more recent additions to this list.

It has often been said that our planet is surrounded by an electric plasma and indeed that the larger part of the Universe is in the plasma state. Yet before we can examine whether this is not merely an exaggeration, intended to make plasma physics look important, it is necessary to enquire into the nature of the plasma state. The kinds of plasma we are going to discuss are usually in the form of a gas and are electrical in origin (figure 1.1).

First we shall try to answer the question "what is a plasma?" and give a simple explanation of the word. A plasma is electrically energized matter in a gaseous state. In general it consists of three components: electrically neutral gas molecules; charged particles in the form of positive ions, negative ions and electrons; and quanta of electromagnetic radiation (photons) permeating the plasma-filled space. The molecules, of which there are often a large number of species, can be either in their lowest ('ground') energy state or in a higher ('excited') energy state of the rotational, vibrational and electronic kind, each molecule consisting of one or more atoms. Positive ions can carry one or more units of charge and, again, may be in their ground energy state or an excited state, whereas negative ions carry only one single charge. Though in most cases such plasmas are caused by an electric discharge in a gas, they may also be produced in other ways: by sudden or continuous heating of matter to high temperatures, by intense laser radiation and by chemical processes. We shall see later on how this can be practically achieved.

From what has been said above, the question arises as to whether the

1

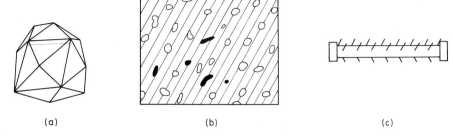

Figure 1.1. Three kinds of plasma.
(*a*) A green variety of quartz.
(*b*) The coagulable part of blood.
(*c*) The electrically conducting gas at low pressure frequently emitting 'cold light'.

structure of a plasma might perhaps be akin to that of an ordinary electrolyte. What is the difference between the two? Firstly, most of the so-called electrolytes are liquids and only a few are solids. Secondly, in electrolytes free electrons and photons are absent; the only charged particles present are negative and positive ions. Finally, apart from certain exceptions, all the electrolyte particles, whether neutral or charged, are in the ground state because they collide frequently with one another.

The somewhat ambiguous term 'plasma' was introduced by Langmuir in 1927. In the course of his studies of electric discharges in mercury vapour at low pressure he found that the ionized gas, which formed a cold luminous column in a long glass tube, showed uniform electric and optical properties along its whole length. From that he concluded, as Crookes had done in 1870, that he was dealing with a particular state of matter. It deserves to be given a special name which conveys that the various constituent particles are moulded together: thus the Greek word *plasso* (to shape or mould). As we shall show later on, his foresight has been confirmed.

Let us approach this subject from another angle. Everyone has observed lightning, and those who have lived sufficiently far north or south must also have seen the aurora (borealis or australis). The 'long sparks' between cloud and ground, and the strange diffuse lights emanating from the cloudless sky are visible evidence of the presence of intense and weak plasmas. Some plasmas emit visible light by virtue of the electric energy supplied to a discharge, for example, sodium or mercury vapour lamps, which also contain a rare gas at low pressure to facilitate starting. Other familiar plasma (cold light) sources are neon lamps and signs. In contrast, high-current electric arcs used for welding in atmospheric air or other gases develop very hot spots at the electrodes. Such bright and noisy arcs are used for joining rails or when the 'gliding shoe' of an electric train passes a connecting section, thereby opening the circuit temporarily; welding arcs are widely used on building sites and shipyards for fabricating metal structures.

Plasma physics is not limited to electric discharges in gases. The application to chemical processes has been studied in recent years with great vigour. It has long been known that certain properties of oils can be impoved by treating them with electric corona-like discharges, which in air or oxygen also produce ozone used for purifying water. Atomic hydrogen generated in hot torch discharges is also used in welding and cutting. Reactions that occur in internal combustion engines and in flames are usually accompanied by ionization and excitation, processes which do not necessarily take place in the hottest part but rather in the chemically most active part of a flame gas. Flame plasmas are another field of plasma study.

Other practical applications include the gas laser, in which a plasma is the active medium. In general, a combination of several gases is used whereby the radiation produced controls the emission of light from certain excited atomic or molecular levels. With the help of an 'optical cavity' an intense beam of coherent monochromatic light is emitted by the plasma. Again, relatively weak glow discharges are employed either for keeping a d.c. voltage constant or replacing mechanical switches in low current circuits. Still weaker discharges occur in gas-filled counters: they develop when a single particle or a single photon enters a chamber, and quenches itself automatically so as to be ready for the next arrival. More powerful temporary discharges are found in spark counters used in cosmic work: the luminous trace which persists when a cosmic particle from outer space has passed through the counter represents its path and is photographed. High-energy particle accelerators are fed from negative or positive ion sources. These are often high-frequency discharge plasmas from which slow ions are drawn into the beam accelerator. In standard electric switch gear, arcs develop between separating contacts before the current is interrupted. These are examples of the applications of discharge and plasma physics, some of which will be discussed in later chapters.

We can listen to radio broadcasts from distant stations because Nature has provided a medium which enables us to receive signals from these far corners; though sometimes faint, they are frequently accompanied by electric disturbances from space, particularly during the summer or periods of unstable weather. This medium, a weak plasma maintained by the Sun's activity surrounding the Earth about 100 km or more above its surface, is called the ionosphere. Owing to its presence we can keep up communication between distant parts of the globe without recourse to satellites. Signals from a transmitter are reflected by this plasma shell, sometimes more than once, provided the signal beam carries sufficient power and has the appropriate frequency to be reflected down to Earth. However, we must not conclude that signals originating from sources at heights above the ionosphere or other planets cannot penetrate the ionosphere. This is shown by the reception of signals from radio stars and space craft.

A field of plasma physics which may be of great benefit is thermonuclear fusion—an exothermal reaction of immense potentiality. Unlike the hydrogen bomb, here the fusion energy is released at a carefully controlled slow rate. In fusion studies plasma physics and nuclear physics overlap. When a pair of light nuclei (deuterium $D = {}^2H$ and tritium $T = {}^3H$) are brought closely together by colliding at high speed, they fuse, a reaction followed by nuclear disintegration into neutrons and other nuclei. Fusion is achieved by heating a rarefied gas mixture of D and T with an electric discharge. The plasma is kept away from the container walls by a magnetic field, until it reaches temperatures of 10^6–10^8 K. Alternatively, giant lasers are being operated now to fuse D and T together. Yet after more than 25 years of intense research in various countries we still seem to be a long way from our scientific goal and still further from the power station of the future housing a fusion reactor.

There is another venture in which the study of plasma physics may possibly bring success: that is by producing electrical energy through driving a hot plasma across a magnetic field, which is analogous to replacing the rotating copper rods or wires in an ordinary generator or dynamo by a flow of gas of high electrical conductivity. Alas, the results up to now are not what we hoped for, though such magnetohydrodynamic generators have been used to provide quite large electric power pulses for a very short period of time at low efficiency.

CHAPTER 2

nature, structure, state and generation of plasma particles

2.1. Neutral atoms and molecules

The mass of an atom is essentially concentrated in its nucleus or core, which is usually composed of protons and neutrons (exceptions are hydrogen and positronium). An early model of the neutral atom is that of Niels Bohr. This consists of negative electrons, of mass small compared with that of the positive nucleus, which move in selected circular or elliptical orbits around the nucleus. By knowing the number of electrons in the orbit and the type of orbit (energy level), the potential energy (state of excitation) of the atom can be found. When a bound electron in a lower level is moved to a higher empty level, i.e., from an inner to a larger outer shell, the atom's excitation energy is raised. However, the number of electrons that can occupy a shell is restricted, since Pauli's exclusion principle stipulates that each atomic electron differs from all the others by at least one of its quantum numbers, which characterizes its property in the shell structure. Similarly, members of a large family are identified by their family name plus at least one different forename. In terms of this model, at least four numbers are needed to describe an atomic state: the principal quantum number n, describing the electron's orbital angular momentum, the azimuthal quantum number l, describing the ellipticity of the electron's orbit, the orbital magnetic quantum number m_l and the electron spin quantum number m_s.

After half a century this simple picture has been replaced by a more sophisticated concept based on quantum mechanics (see, e.g., *Elementary Quantum Mechanics* by N. F. Mott, Wykeham Science Series No. 22). However, what has really been superseded is the idea of an electron literally rotating like a planet around the Sun in an orbit that can be traced. The quantum numbers, the energy states and the exclusion principle remain. Though one speaks now of 'orbitals' and 'energy levels', not of orbits, and thereby strengthens our belief in the wave character of the electron, an outline picture of the Bohr atom is still very valuable. It is used in the same way as are the particles in the classical kinetic theory of gases, where molecules are described as massive elastic spheres, or in chemistry where molecules are depicted as spherical atoms held in a scaffolding of bonds.

2.2. Positive ions

On the left of figure 2.1 is the simplest atom, hydrogen, with the proton at its centre and a single electron in the atom's ground state, i.e., in an orbit of least radius (the Bohr radius $a_0 = 5 \cdot 3 \times 10^{-11}$ m). The radius of the proton is about $1 \cdot 2 \times 10^{-15}$ m, and the classical radius of the electron, $r_e = 2 \cdot 8 \times 10^{-15}$ m, so the intervening space is apparently empty, i.e., free of 'particles' with inertial mass. On the right of figure 2.1 a single proton is shown which can be obtained by removing the electron from a neutral hydrogen atom. The spatial distribution (topology) of the electric field is of interest. There is the 'closed field' structure of the neutral hydrogen atom, where the circling electron charge prevents the electric field of the nuclear proton from penetrating much further than the innermost orbit. In contrast, the proton's (H^+) electric field with its radially diverging field lines, ending at the electron nearby or at infinity, exemplifies an 'open structure'. In larger neutral atoms the positive charge at the centre is surrounded by electrons circulating in various orbits whose planes are differently oriented. However, here the electric field at larger distances from the nucleus is gradually screened off by the electrons. Thus, the stray field outside the large atom has a structure which is more open the greater its size. The increase in polarizability with increasing atomic radius or atomic number confirms this view (figure 2.5, p. 12).

The more strongly bound the outermost electron is to an atom, the larger is the energy needed to excite it or to ionize it. Another related fundamental atomic property, affecting the 'shape' of particles and the path of moving charges, is their electric polarizability. It is a measure of the elastic deformation of the electronic charge distribution around the nucleus. Induced polarization occurs when an electric field E acts on the atom's electron cloud and displaces its centre by a distance δa relative to the nucleus. This results in an induced electric dipole moment

$$M_e = q\delta a = \alpha E$$

where α is called the polarizability and q the total electronic charge.

The order of magnitude of α is easily estimated by balancing the internal against the applied (distorting) field. Since the internal field E_0 at the innermost orbit, a_0, is $e/4\pi\varepsilon_0 a_0^2$ (where e is the electron charge, $1 \cdot 6 \times 10^{-19}$ C, and ε_0 is the vacuum permittivity, $8 \cdot 85 \times 10^{-12}$ F m^{-1}) and the applied field E changes a_0 by $\delta a \ll a_0$, $E/E_0 \simeq 4\delta a/a_0$, and $\alpha \simeq 0 \cdot 25 a_0^3 > 0 \cdot 25 \times 10^{-30}$ m^3 in agreement with observations (table 2.1). Figure 2.6 shows how α depends on the atomic number Z, i.e., the number of positive nuclear charges; it confirms that the values of α of rare gases are associated with minima, due to their closed shells, and those of

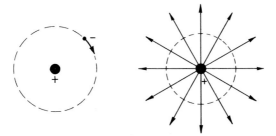

Figure 2.1. H atom and H$^+$ ion (proton).
  Dashes = electron orbit; arrows = electric field.

alkali atoms with maxima, owing to their open structure causing strong chemical activity. Excited atoms, positive ions and molecules have low values of α. Polarizability is of considerable interest in ion-cluster formation, mobility theory, collision theory and kindred subjects.

Bohr's picture of the atom requires various amendments but only one, namely the assumed 'point charge' of the electron, will be discussed. In accordance with the uncertainty principle, the electronic charge must be taken as being 'smeared out', that is, it has a radial and azimuthal distribution; the former has been deduced from the observed intensity distribution of X-rays that are scattered by orbital electrons. Figure 2.2(a) shows the results for He and Ne. The maximum intensity for He is seen to be at the radius of its orbit $n = 1$, while the maxima for Ne are at about the first and second orbital radii ($n = 1$, $n = 2$) just as quantum mechanics predicts. It is thus impossible to locate precisely the position of the electron at any instant. Heisenberg's uncertainty principle applied to circular motion states that the uncertainty in the angular momentum ΔJ and the uncertainty in the angular co-ordinate $\Delta \phi$ must satisfy the inequality $\Delta J \times \Delta \phi > h/2\pi$ ($h \simeq 6 \times 10^{-34}$ J s). If we take $\Delta J = m_e \times \Delta(v \times r)$, where v is the electron velocity corresponding to the ground-state energy ($n = 1$) of H (which works out to be 2×10^6 m s^{-1}) and the electron mass $m_e = 9 \times 10^{-28}$ g, then the uncertainty Δr comes to $r \simeq 5 \times 10^{-11}$ m which is the same as the Bohr radius a_0. We conclude that this forecast is consistent with experiment. The radial and azimuthal charge distribution helps to answer the question of why the revolving charge does not emit radiation as demanded by classical (point-charge) theory: the elements of radiation emitted by the revolving element of the electron cloud annul each other by interference.

The ellipticity of discrete electron orbits is replaced in quantum mechanics by boundary surfaces (orbitals) which indicate the charge distribution in given

Table 2.1. Polarizability α of neutral unexcited atoms (in 10^{-24} cm^3).

Atom	He	Ne	Ar	Kr	Xe	H	Be	O	Hg	Li	Na	K	Rb	Cs
α	0·21	0·4	1·65	2·5	4·04	0·67	9·3	0·15	5·2	20	27	38	50	50

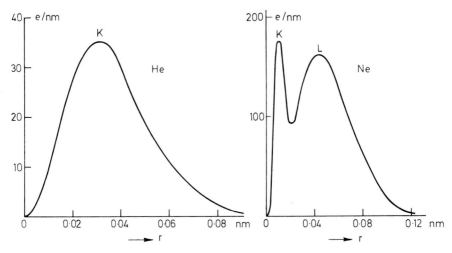

Figure 2.2. (*a*) Radial distribution of electron charge (derived from X-ray scattering) of He and Ne. (After E. O. Wollan, 1931, *Physical Review* **38**, 15.)
Maxima = Bohr orbits.

directions for electrons of the s($l=0$), p($l=1$), d($l=2$) and f($l=3$) type and corresponding to the azimuthal quantum numbers $l=0$ to 3. We note that s electron orbitals are spherically symmetrical (figure 2.2(*b*), bottom), p orbitals (figure 2.2(*b*), top) look like two axially symmetrical balloons, and so on. A full

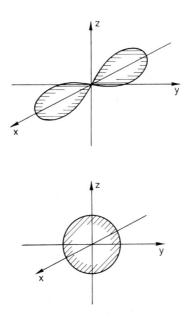

Figure 2.2. (*b*) s and p orbitals: bottom and top, respectively.

description of an atomic electron still requires at least four quantum numbers: besides n and l, the magnetic quantum numbers m_l and m_s are the components of the angular momentum l and the spin vector s in the direction of an applied external magnetic field B. Their magnitudes depend of course on the orientation of the vectors l and s with respect to the direction of B. l precesses about B at discrete angles, whereas s can only have one of two positions, either parallel or anti-parallel to B.

So far we have referred to charged and uncharged atomic particles, which are for $n = 1$ all we find in rare gases and metal vapours, but diatomic and polyatomic molecular positive ions are of considerable interest. Positive molecular ions can be singly or multiply charged. (Atomic ions can be completely stripped of electrons and some that have lost more than 30 electrons have in fact been observed spectroscopically, but a completely ionized atom, say He^{2+}, can neither be excited nor emit light.)

We know that ordinary hydrogen gas consists of diatomic molecules in their electronic ground state. Their two H atoms are not held together by ionic bonds, as are the Na^+ and Cl^- ions in a sodium chloride lattice, but by 'exchange forces', the so-called covalent bond, which arises from the two electrons being shared between the atoms while circling around the two nuclei in a figure of eight. Thus, when the negative charges, i.e., their distributions, are in the space between the two positive (repelling) nuclei, an attractive force acts on them temporarily, but when they are outside the nuclei a repulsive force drives them apart (figure 2.11, p. 20). The average force keeps the atoms bound together a certain distance apart. If the molecule is electronically excited the electrons travel along a distorted egg-shaped orbit (approaching the figure of eight but never crossing the centre between the nuclei) whereupon the atoms separate within a time of $\sim 10^{-13}$ s, the vibration period. When one electron is removed from the molecule a molecular ion H_2^+ results. Now only one electron binds the nuclear pair and the interatomic distance of H_2^+ should be larger than that of H_2, and this has been

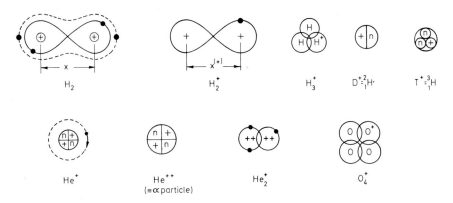

Figure 2.3. H_2 molecule and a selection of atomic and molecular positive ions.

confirmed. Obviously, the removal of the remaining electron will produce two protons which repel each other. However, because of the open field structure, an H_2^+ ion can attach a neutral H atom, so forming a stable H_3^+ ion (figure 2.3); H_4^+ and H_5^+ and other cluster ions have also been observed.

Of particular interest in plasma physics and its application to nuclear fusion are deuterium ($D = {}_1^2H$) and tritium ($T = {}_1^3H$), hydrogen isotopes with one and two neutrons in the nucleus, respectively (figure 2.3). The corresponding positive ions are D^+ and T^+. Helium, the lightest rare gas atom, whose nucleus ${}_2^4He$ consists usually of two protons and two neutrons (or two protons and one neutron in case of its rare isotope ${}_2^3He$), can form the singly charged ${}_2^4He^+$ ion or the doubly charged ion He^{2+}, the latter being an α particle which is emitted with large kinetic energy in radioactive decay.

2.3. Negative ions

Consider atomic and molecular negative ions which form when electrons become attached to neutral particles. We can understand electron affinity in terms of the degree of 'openness' of an atomic or molecular field (figure 2.1). If the electric field of an H atom did not extend beyond the Bohr orbit a_0, an electron approaching the atomic structure would move in a zero-stray field and attachment would rarely occur. However, though a stray field decreases rapidly with the distance from the atom's centre, it has a finite value beyond a_0 and attracts a slowly approaching electron. When the free electron is near the atom at a_0, the other bound electron is slightly displaced outwards, and the structure opens, thereby strengthening the force on the 'supernumerary' electron. Thus, H^- is born, which has two electrons orbiting around the nucleus, but another electron is not attachable. No doubly charged negative ions have been observed up to now, nor have H_2^- ions.

Of particular interest is the so-called helium negative ion (figure 2.4). This is a misnomer. An ordinary helium atom neither attaches an electron nor acts chemically. However, an electron can become attached to an excited helium atom He^*. This can be understood by invoking again our argument based on the topology of atomic fields. In He^* one of the electrons occupies its innermost orbit and the other an orbit of a larger radius. Hence the 'excited structure'. The stray field extends now to larger distances, which facilitates electron attachment which is weak and only temporary. A more correct notation of this ion is $(He^*)^-$. Rare gas–halide (excited) molecules appear in lasers. Dissociative attachment, $H_2 + e \rightarrow H + H^-$ or $H_2 + e \rightarrow H + (H^*)^-$, $n = 2$, also yields H^- ions.

In a sense this species is similar to the short-lived negative molecular nitrogen ion $(N_2^*)^-$. Here a free electron, initially caught in the molecular field, seems to orbit it many times before detaching spontaneously. This 'temporary' ion revealed itself over 50 years ago as a large peak in the measured electron

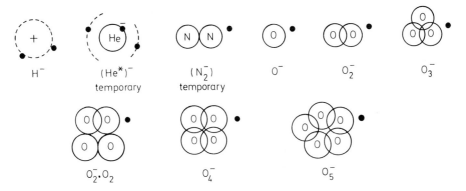

Figure 2.4. Temporary and stable negative atomic and molecular ions.

scattering cross-section of N_2 (see Chapter 3), but its formation has only recently been explained.

Another interesting example of the large variety of negative ions is offered by oxygen (figure 2.4) which has atomic, diatomic, triatomic and tetratomic negative molecular ions (O^-, O_2^-, O_3^- and O_4^-). It is not surprising to find that the 'affinity energy' needed to detach the extraneous electron decreases with the increasing number of atoms per particle. The likelihood of producing atomic negative ions can also be explained in terms of the atomic polarizability (figure 2.5) and thus of the position of the element in the periodic table. Halogen atoms and molecules may readily attach an electron (e.g., Cl^-, Cl_2^-) and so do radicals such as OH^- and CN^-.

Larger positive and negative ions can result from the attachment of a small ion to a large molecule or complex, provided the attractive forces between them are sufficiently strong to survive thermal collisions. The strength of the forces and potentials depends on the (induced) electric polarizability of the large molecule that will be discussed in Chapter 3.

2.4. Ionization and excitation of atoms and the Periodic Table

A number of physical and chemical properties show a periodic variation with the atomic number Z. In the Periodic Table (p. x) the elements are grouped together in such a way that those in the same column have similar chemical, electrical, thermal and other properties. Does this periodicity also apply as regards the energy needed to produce positive ions from neutral atoms? It does indeed.

The noble gases are chemically inert when in their normal (ground) energy state, because each of their shells contains the largest number of electrons permitted by the exclusion principle—hence the closed shell. This may explain

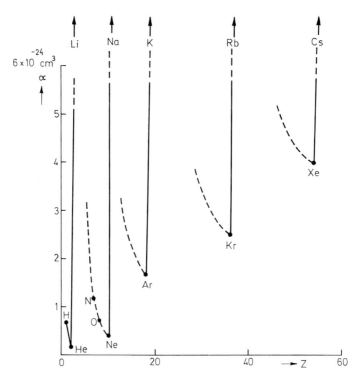

Figure 2.5. Induced (electronic) polarizability α of atoms in the ground state as a function of their atomic number Z.
α in cm^3; induced dipole moment $M = \alpha E = ed$. For $E = 1$ e.s.u. (300 V cm^{-1}) the displacement $d \simeq 10^{-14}$ cm $\ll a_0$. α is larger for excited atoms and negative ions, and smaller for positive ions. α of Li, Na, K, Rb, Cs and Fr is 20, 27, 38, 40, 60 and 60, respectively; of Hg: 10.

why each rare gas atom requires very much more energy to remove a bound electron than one of its neighbours in the Periodic Table (p. x). For example, neon $(Z = 10)$ has two electrons in the first $(n = 1)$ and eight electrons in the second shell $(n = 2)$. It is preceded by fluorine $(Z = 9)$ which has an incomplete second shell of only seven electrons and which readily forms a negative ion F^-. The attached electron does not simply enter the unfilled shell but is loosely held outside it by a field resulting from the nine positive and nine partly rearranged electronic charges. We expect *all* halogens to exhibit similar properties, and this is the case. It is also not surprising that fluorine's neighbour oxygen, as well as the other elements in the same column as oxygen, tend to favour electron attachment. The halogens, the column comprising F to At, are the most strongly electronegative elements.

The alkali metal elements (Li to Fr) are called electropositive because of their readiness to dispose easily of their solitary outermost electron. The removal of this lonely electron is energetically easy, because the nuclear field is well shielded by the electrons in the inner shells. The chemical activity of the alkali metals is therefore exceptionally high.

Let us continue our argument about the tendency to attach or dispose of an electron in a more quantitative way. When a neutral atom is hit by a sufficiently fast electron so that the energy imparted exceeds a critical value, the ionization energy E_i, the atom may lose one electron of the outer shell. This process—ionization by electron collision—can be described symbolically by

$$A + (e + E_k) \rightarrow A^+ + (2e + E_k')$$

where A stands for the atom, e for the electron, A^+ for the positive atomic ion, E_k for kinetic energy of the incoming electron and E_k' for the sum of the kinetic energies of the two free electrons after ionization (the positive ion acquires substantially no kinetic energy).

In order to ionize a neutral atom in the ground state, E_k must be greater than E_i. To free a bound electron from an atom requires an energy transfer exceeding $E_i = eV_i$ where V_i is the ionization potential, i.e., the energy acquired by an electron that moves without colliding from a point at $V=0$ to one at $V=V_i$. This is also termed the 'first ionization potential', since the removal of a second electron from the positive ion requires an additional energy V_{i2}, called the 'second ionization potential'. The electron-volt energy scale is most convenient because V_i of the atoms lies between about 4 and 25 V. The corresponding absolute energy is $1 \cdot 6 \times 10^{-19}$ J V^{-1}. By expressing E_i in eV, we find that E_i and V_i are numerically equal. Thus, ionization energy and ionization potential can be taken as synonymous.

When an atom at rest is ionized by a moderately fast electron, the energy transferred may be larger than its ionization energy. Energy conservation demands that the surplus is partly carried away by the primary electron and partly by the released (secondary) electron, provided the potential (excitation) energy V_{ex} of the atom remains the same. This also holds if the atomic electron, instead of being set free, is only raised to a higher energy level. The excited neutral atom so formed has a finite average life ($<10^{-8}$ s). We shall now show how these two 'critical potentials', V_{ex} and V_i, of the elements are related to the position of an atom in the periodic table.

In figure 2.6 the ionization potential V_i of atoms is plotted against the atomic number from $Z=1$ to $Z=60$. The values of V_i of the rare gases appear as peaks whose magnitudes decrease with increase in the principal quantum number n, which determines both the atomic radius and the position in the column of the periodic table. This is to be expected, since as n increases, an outer electron, about to be removed, is further away from the nucleus; also, the effective field acting on it is smaller because of the stronger shielding. There is a sudden drop of V_i in going from the rare gas to an alkali metal atom, but a gradual increase of V_i towards the next closed shell. All the alkali metals occupy minima on the ionization potential curve whose values decrease as the size of the atom increases; thus, for example, V_i (caesium, Cs) $< V_i$ (lithium, Li).

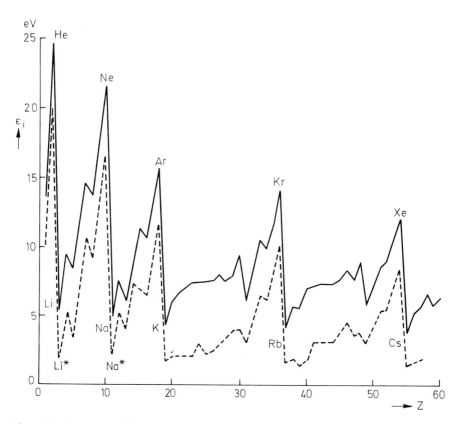

Figure 2.6. Ionization and lowest excitation energies of atoms as a function of Z.

The second, dotted curve in figure 2.6 shows how the lowest excitation potential V_{ex} of an element or the energy E_{ex} in eV producing the lowest excited state depends on the atomic number Z. The general pattern of this curve is the same as that of $V_i = f(Z)$. This we expected, since for every element V_{ex} must be below V_i.

Here we must note that similar rules do not apply for molecules. Consider, for example, the ionization potentials of the homopolar molecules N_2, O_2, H_2. If we were guided by the ionization potentials of the atoms N (14·5 V), O (13·5 V) and H (13·6 V), and of the molecules N_2 (15·5 V) and H_2 (15·4 V), we might expect that V_i for O_2 should exceed 13·5 V; in fact it is only 12·1 V. Neither do the rules apply to the lowest (electronic) excitation potentials of molecules, a subject to be discussed below.

Returning to the atom, its characteristic features are easily derived from the energy-level diagram (figure 2.7). This shows only the principal levels of H, the notation of the various states of excitation and their potential energy, assuming for convenience zero potential energy for the ground state. We observe a wide energy

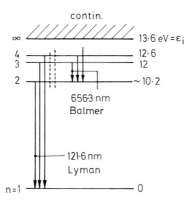

Figure 2.7. Energy level diagram and the different series of the H atom.
Lower state $n = 1, 2, 3, 4$ corresponds to Lyman, Balmer, Paschen, Brackett series.

gap between $n = 1$ and the lowest excited state $n = 2$ at $10 \cdot 2$ eV ($\sim 0 \cdot 75\ V_i$). Compare the corresponding gaps of He ($\sim 0 \cdot 8\ V_i$) and Li ($\sim 0 \cdot 3\ V_i$). The higher levels ($n > 2$) are separated by shrinking intervals, until finally the discrete levels converge to a limit (V_i) beyond which lies the continuum. In this region the electron, now separated from its atom, is able to have any value of energy.

The ionization energy of atomic hydrogen can be derived from the equations of total energy, force and quantization of angular momentum of the electron ($\hbar = h/2\pi$; 1 eV $= 1 \cdot 6 \times 10^{-19}$ J)

$$eV_i = -\frac{Ze^2}{(4\pi\varepsilon_0)r} + \frac{mv^2}{2}; \quad \frac{Ze^2}{(4\pi\varepsilon_0)r^2} = \frac{mv^2}{r}; \quad mvr = n\hbar \tag{2.1}$$

Eliminate v from equation 2.1, find r and solve for eV_i; thus $r \propto n^2/Z$, $R_H = 1 \cdot 1 \times 10^7$ m^{-1}

$$|eV_i| = \frac{me^4}{2\varepsilon_0^2 h^2}\left(\frac{Z}{n}\right)^2 = hcR_H\left(\frac{Z}{n}\right)^2 \tag{2.2}$$

where R_H is the Rydberg number for H. Inserting the values of the constants from table 2.1, the ionization energy of H, with $Z = 1$, $n = 1$, is $eV_i = 13 \cdot 6$ eV $= hcR_H = 2 \cdot 18 \times 10^{-18}$ J. The excitation energies of a hydrogen atom to the level n are therefore

$$eV_{ex} = 13 \cdot 6\left(1 - \frac{1}{n^2}\right) \text{eV} = 2 \cdot 18 \times 10^{-18}\left(1 - \frac{1}{n^2}\right) \text{J} \tag{2.3}$$

where $n = 2, 3, 4 \ldots$ since for $n = 2$, $eV_{ex} = 10 \cdot 2$ eV, the ionization of H*($n = 2$) requires $V_i^* = 3 \cdot 4$ eV (figure 2.7). The energy of the spontaneously emitted photon (quantum $h\nu$) of the reaction

$$H^* \rightarrow H + h\nu$$

is $h\nu = 10\cdot2$ eV and the corresponding wavelength $\lambda = c/\nu$, ν being the frequency. Hence, with the velocity of light $c = 3 \times 10^8$ m s^{-1}

$$\lambda = \frac{hc}{eV_{ex}} = \frac{6\cdot63 \times 10^{-34} \times 3 \times 10^8}{10\cdot2 \times 1\cdot6 \times 10^{-19}} = 1\cdot21 \times 10^{-7} \text{ m} = 121 \text{ nm}$$

(1210 Å) which lies in the far ultraviolet. The reverse process is called photoexcitation.

If a positive ion has Z positive nuclear charges and one electron only (He$^+$, Li^{2+}, Be^{3+}, B^{4+}, etc.) its ionization energy V_i is again easily found because the equations 2.1 hold for two point charges only. Since $(eV_i')\propto Z^2$, the process He$^+ \rightarrow$ He^{2+} requires that an energy $(Z=2)$ of $4 \times 13\cdot6 = 54\cdot4$ eV is transferred to He$^+$. Thus, if He^{2+} is produced from neutral He $(V_i = 24\cdot5$ eV) a total energy of $78\cdot9$ eV is needed.

Consider now an H atom whose electron in the level n has made a transition to m. The wavelength of the photon emitted when returning to n is given by

$$\frac{1}{\lambda} = R_H \left(\frac{1}{n^2} - \frac{1}{m^2} \right) \tag{2.4}$$

n being the lower and m the higher quantum number. The group of hydrogen emission lines whose lower levels are $n=1$, 2 and 3 are called the Lyman series, the Balmer series and the Paschen series, respectively (figure 2.7).

Such simple calculations as equation 2.1 fail if the initial atomic system contains two or more electrons. Here the concept of 'screening' has to be introduced. Take Li with two $n=1$ electrons, one $n=2$ electron and $Z=3$. Guess the energy necessary to remove the outer electron. If the two electrons in the inner closed shell were at the centre, the net charge there would be as if $Z=1$ and $V_i \propto (Z^2/n^2) = (1/n^2)$, while if we neglected shielding by the two electrons altogether, $Z=3$ and $V_i \propto (9/n^2)$. Somewhere between these limits the 'effective' value of n must lie, certainly nearer $1/n^2$. Physically, the two electrons partly shield the nuclear field from the single outer electron and facilitate its removal. Allowing for shielding by using an effective value of the principal quantum number only and keeping $Z=3$, we must raise n and thus $1 < n_{eff} < 2$, because the outer electron attracts the nucleus and repels the two inner electrons so that a dipole field develops. The corresponding potential varies thus with $1/r^4$, and when superimposed on the Coulomb potential $e^2/4\pi\varepsilon_0 r$ forces the outer electron to move in an orbit of high eccentricity. Hence, $n_{eff} = n - n'$ and $m_{eff} = m - m'$ should differ considerably from an integral number where the 'quantum defects' n' and m' (between 0 and 1) allow for screening. For all alkali atoms a relation similar to equation 2.4 is obtained for the energy level differences, i.e., the quantum emitted is

$$h\nu = hcR \left[\frac{1}{(n-n')^2} - \frac{1}{(m-m')^2} \right] \tag{2.5}$$

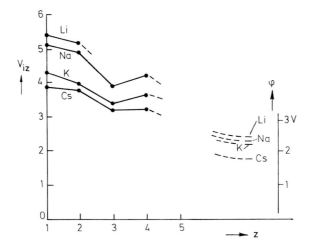

Figure 2.8. Ionization energy of mono- and polyatomic molecules as a function of the number of atoms Z.
Curves indicate the transition from ionization energies eV_i to work functions φ.

Screening and quantum defects are best known for sodium (Na) which has two closed inner shells. Thus, analogous to equation 2.2, $eV_i = hcR_H/(n-n')^2$. For Na, $n=3$ and $n'=1\cdot37$; hence, $V_i = 13\cdot6/(1\cdot63^2)=5\cdot1$ V, showing that here n_{eff} is about one half of n (figure 2.8). Obviously, screening is strongest for electrons in levels outside closed shells and when Z is large.

With increasing Z the shells 1, 2 and 3 are gradually filled with up to two, eight and 18 electrons before the next higher shell begins to be occupied, but this is no longer true when Z is greater than 20 (figure 2.9).

Finally, the effect of the interaction between the orbital magnetic moment **l** due to the motion of the electron about the nucleus and the spin magnetic moment **s**, associated with its own spin ($s=0\cdot5$), has to be considered. The vector sum $\mathbf{j}=\mathbf{l}+\mathbf{s}$ of the two moments is consistent with strong interactions between *all* orbital vectors and between *all* spins, whereas orbital and spin vectors of the same electron interact only weakly or not at all. He (figure 3.8, p. 43) emits two series of lines: one in which the electron spins are antiparallel and one in which they are parallel, forming the singlet and triplet level systems respectively, which do not interact. Thus, no line can be emitted which starts from a level in one and ends at a level in the other system. The three-fold tripled states of He arise for the following reasons. The sum **J** of all **l** vectors (**L**) and all **s** vectors (**S**) is unity. Since $\mathbf{L}+\mathbf{S}=\mathbf{J}$, we obtain for **J** the values $1+1$, $1-1$, and 1, i.e., we have $\mathbf{J}=2, 0, 1$. Note that when $\mathbf{J}=1$, the components **L**, **S** form an equilateral triangle. We conclude that a P state ($\mathbf{L}=1$) thus splits into $2\mathbf{S}+1=3$ levels. The alkali atoms with their closed inner shells have one electron in the following outer one. Thus, since $s=0\cdot5$, $2\mathbf{S}+1=2$ and thus alkalis emit doublets. We remember the Na

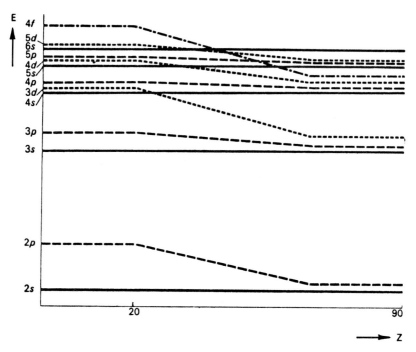

Figure 2.9. Energy (in arbitrary units) of the various atomic sub-levels as a function of the atomic number Z.
(M. F. C. Ladd and W. H. Lee, 1969, *Modern Physical Chemistry*. Harmondsworth, England: Penguin Books.)
Note that for s levels E is independent whereas for p, d and f levels E depends on Z when $Z > 20$.

doublet, the close pair of yellow lines at 599·0 and 599·6 nm, and the doublet structure of H.

The alkalis have no 'metastable' states. These are energy states from which transitions to the ground state (accompanied by emission of a quantum) are forbidden. Metastable atoms (rare gas, mercury (Hg), N, O) keep their potential energy for a time interval which is long ($> 10^{-3}$ s) compared with the free life of a radiating excited state ($< 10^{-8}$ s), provided only that atoms are not perturbed by collisions or external fields. Strictly speaking, it is necessary to distinguish between weak and strong metastability. Resonance states are those which, when excited, emit resonance radiation caused by transitions with the ground state and derive their name from the strong absorption in their own gas.

2.5. Molecular states and electric dipoles

Molecules are structurally different from atoms, as is shown by their spectra; the bonding energy which keeps the atoms together is often of the same order as the first excitation energy of an individual atom. Moreover, a diatomic molecule (dumb-bell model) can rotate about two axes (the third contributes negligible rotational energy) and vibrates along the bond. Classical quantum theory gives the values of a rotational quantum of the order 10^{-3} to 10^{-2} eV, and of a vibrational quantum $<0 \cdot 5$ eV. A typical potential energy $(E = f(r))$ diagram is shown in figure 2.10 for the principal lowest electronic states of $H_2 (^1\Sigma_g, ^3\Sigma_u)$ and some vibrational levels $v = 0$, 1, 2, and so on as a function of r where r is the instantaneous interatomic distance. These vibrations are naturally asymmetrical, shown by the dash–dotted line which goes through the centre $\bar{r}$ of the classical vibrational amplitudes; only the lower vibrational levels v approximate to those of a harmonic oscillator. The oscillation amplitude is small $(r_c - r_B < \bar{r}_0)$ compared with the average radius $\bar{r}$ of a molecule at ordinary temperatures T, i.e., for small values of v; but when T and v are raised the amplitude approaches infinity and the molecule dissociates. The vertical distance between $v = 0$ and $v = \infty$ represents

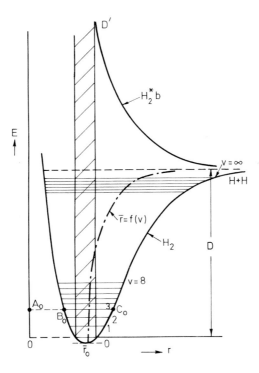

Figure 2.10. Potential energy diagram of the H_2 molecule in the lowest attractive ($=$ ground) state and H_2^* in the lowest repulsive (electronically excited) state, i.e., $E = f(r)$.
$r =$ instantaneous value of interatomic separation; $\bar{r} =$ average distance between atomic centres; $v =$ vibrational quantum number $(0, 1, 2, \ldots \infty)$; shaded region $=$ Franck–Condon range; $D =$ dissociation energy; $B_0, C_0 =$ smallest and largest nuclear separation, respectively. The zero point energy $E = 0$ is indicated by horizontal dashes near $\bar{r}_0$.

the thermal dissociation energy E_D; for H_2, $E_D = 4 \cdot 5$ eV ($\sim$100 kcal mol^{-1}). The anharmonicity of the vibration causes a shift of its atomic centres to larger r_0 in analogy to the thermal expansion of solids. The atoms of H_2 in the electronic ground state ($^1\Sigma_g$) are held together by 'covalent' bonding (see figure 2.3 and text on p. 9).

So far we have dealt only with the electronic ground (singlet) state ($^1\Sigma_g$) of H_2 when the two electron spin vectors are anti-parallel. However, the first and lowest electronically excited (triplet) state of H_2 ($^3\Sigma_u$) with its pair of parallel electron spins (figure 2.11) is repulsive since $dE/dx < 0$ and thus the two atoms are driven apart because the electrons, instead of moving through the molecule's centre, move along a path wound around both nuclei (see figure 2.3). Therefore, a collision between an electron and an ordinary H_2 ($^1\Sigma_g$) molecule that would induce a transition from the $^1\Sigma_g$ to the $^3\Sigma_u$ state, leading to dissociation, would occur along a 'reaction path' different from that for thermal dissociation, when the molecule passes through all levels of v up to infinity. The reason is that when momentum is exchanged between the electron and the molecule (mass ratio $\sim 1:4000$), the latter receives a very short pulse during which r remains constant. An electron of energy about 10 eV has a velocity of about

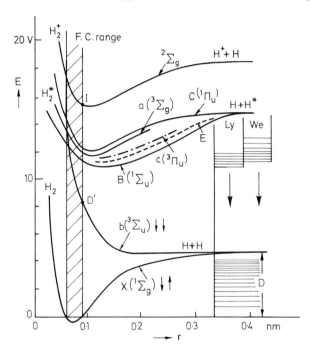

Figure 2.11. Potential energy of a hydrogen molecule in higher excited states (H_2^*) and in the ionized state (H_2^+).

D' = dissociation by electron impact; D = thermal dissociation energy; Ly and We = upper vibrational levels of Lyman and Werner bands respectively; I = ionization energy of H_2.

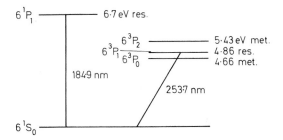

Figure 2.12. Resonance and metastable triplet states and the singlet resonance state of the Hg atom. The resonance line $^3P_1 \to 253 \cdot 7$ nm, an intercombination line, results from the strong spin–orbit coupling, in contrast to the He triplet states (zero coupling).

2×10^6 m s^{-1} and thus 'interacts' with a molecule of radius $0 \cdot 1$ nm for a time Δt of less than 10^{-16} s. Yet the vibrational period, over 10^{-14} s, is very much larger, so the 'reaction or transition path' is a vertical line (figure 2.11), akin to the 'Franck–Condon rule'. Hence, dissociation by electron collision from the $(^1\Sigma_g)_{v=0}$ state can only occur within a region bounded by the two vertical lines which indicate the most likely transition paths from the $^1\Sigma_g$ to the $^3\Sigma_u$ state of H_2^*. Since the right-hand vertical line intersects this (repulsive) curve at a lower energy than the left line, we conclude that about $8 \cdot 8$ eV marks the onset of dissociation; this has been experimentally confirmed. Since the bond energy of H_2 is $4 \cdot 5$ eV, at dissociation each atom carries away $0 \cdot 5 \, (8 \cdot 8 - 4 \cdot 5) = 2 \cdot 15$ eV kinetic energy. The energy surplus cannot be converted into potential energy because hydrogen has no excited state below 10 eV (figure 2.11). This process is associated with a change in electron spin (see the change in superscript from one to three); therefore, one of the bound electrons in H_2 is 'exchanged' with the free incoming electron whose spin has no specified direction. (Dissociation by electron collision can produce 'polar attachment' such as $H_2 + e \to H^+ + H^-$ when the electron energy > 20 eV.) The departing electron carries away energy in excess of $8 \cdot 8$ eV. The principles explained here apply to transitions between electronic (and vibrational) states of molecules, production of metastables, chemical changes or ionization processes. However, transitions caused by absorption of quanta obey different laws, as will be seen in Section 2.6.

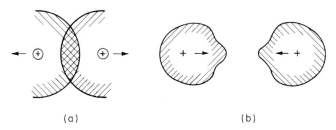

(a)   (b)

Figure 2.13. (*a*) Weak, repulsive van der Waals forces caused by overlap of electron clouds; (*b*) attracting interaction between an electron cloud and a neighbouring nucleus.

Besides the covalent and ionic bonding (typical of, say, H_2 and NaCl, respectively) which is associated with energies of 5–10 eV, and repulsive states (such as $^3\Sigma_u$ of H_2) with energies of similar magnitude but opposite sign, there exist weakly attractive molecular states with energies E_D of about 0·01–0·1 eV as found in Hg_2, Cu_2 and Cd_2. Such molecules are held together by the 'van der Waals' forces $\propto r^{-6}$ (figure 2.13) which are always present. These originate from the electric polarization which occurs when two neutral molecules approach each other, thereby suffering mutual deformation. Their stray fields slightly displace the electron clouds, particularly those in the outer shell, with respect to the positive nucleus, thus producing a weak electric dipole or multipole giving rise to attraction. Quantum theory ascribes this to interaction between dipoles oscillating with electronic frequencies due to zero-point motion of molecules. When they come still closer the outer electron clouds change their shape further. Since the electrons interact strongly, these spatial distributions, where they 'touch', are flattened and strong repulsion develops (figure 2.14). Such simultaneous attraction and repulsion often causes a shallow potential minimum and leads to the formation of weakly bound molecules which partially dissociate at room temperature (kT order 0·03 eV); van der Waals forces are of great importance in the physics of compressed 'real' gases and clusters.

Molecules of like atoms, because of their symmetrical charge distribution, do not form permanent electric dipoles, but those with unlike atoms have a permanent dipole moment in a direction of the internuclear axis because their positive and negative charge centres do not coincide. In contrast to homopolar molecules, they can absorb and emit infrared radiation, since a change in the rotational state, from J'' up to J' or vice versa, is accompanied by absorption or emission of one rotational quantum. The magnitude of a permanent molecular

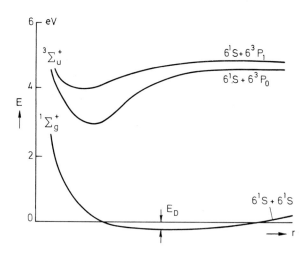

Figure 2.14. Potential energy E of Hg_2, a molecule with a shallow minimum $E_D \sim 0·06$ eV in the ground state.

Table 2.2. Permanent electric dipole moment m_D in Debye units. 1 unit $= 3 \cdot 336 \times 10^{-30}$ C m ($= 1 \times 10^{-18}$ e.s.u.). $m_D = 0$ for CH_4, C_6H_6, CCl_4, CO_2, etc.

Molecule	HI	HBr	HCl	H_2O	H_2O_2	CsF	CsCl	CsI	CO	NO	NH_3	N_2H_4
m_D	0·4	0·8	1·0	1·84	2·2	7·9	10·4	12·1	0·1	0·12	1·4	1·8

dipole moment is of the order $m_D = qa_0 \simeq 3 \times 10^{-30}$ C m for a displacement a_0. H_2O approaches this value of m_D. Table 2.2 gives values of m_D in 'Debye units' for various molecules (one Debye unit $= 1 \times 10^{-18}$ e.s.u. $= 3 \cdot 34 \times 10^{-30}$ C m). We conclude that for all symmetrical (and homopolar) molecules $m_D = 0$, yet their permanent quadrupole moment, arising from two dipoles forming a T shape, is finite. Excited molecules often show $m_D \neq 0$.

2.6. Change in energy state by absorption of radiation in gases

Photo-excitation

Consider an atomic gas through which monochromatic light is passed and assume that its wavelength could be gradually reduced. When the wavelength reaches a critical value such that

$$h\nu = hc/\lambda = eV_{res} \qquad (2.6)$$

eV_{res} being the lowest radiative state—the (first) resonance state—absorption of energy will occur over a narrow energy range of the order of the line width. In table 2.3 a few resonance potentials and wavelengths are given.

It can be seen that the resonance radiation of the alkalis is in the visible range and extends for other elements down to the far ultraviolet range. The absorption coefficient $\mu = f(\lambda) = q(\lambda)N$, q being the absorption cross-section and N the particle concentration. The light intensity drops with the thickness x of the absorbing gas layer as $I/I_0 = \exp(-\mu x)$. Generally, sharp peaks in μ indicate either the onset of photo-excitation or can be associated with other processes such as dissociation, ionization, etc. The process of absorption of resonance light and its significance will now be illustrated. The Hg resonance line has, at room temperature, a Doppler half-width $\Delta\lambda/2 \sim 1 \times 10^{-4}$ nm at $\lambda = 253 \cdot 7$ nm or $\Delta\lambda/\lambda < 10^{-6}$. When the absorption cross-section q is averaged over the line width, i.e., all its components, its value $q(253 \cdot 7) = 1 \cdot 4 \times 10^{-13}$ cm^2, which is very large indeed. By

Table 2.3. Resonance wavelengths and potentials.

Gas	Li	Na	Cs	He	Ne	Xe	Hg	H_2	N_2
V_{res} (V)	1·85	2·1	1·4	19·8	16·6	8·4	4·86	7·0	6·3
λ(nm)	670	589	894	62·6	74·6	147·5	253·7	177	197

irradiating Hg vapour with resonance light, ordinary (6^1S_0) atoms are lifted to the triplet resonance state $^3P_1$ (figure 2.12)

$$Hg(6^1S_0) + h\nu \rightarrow Hg*(^3P_1)$$

When $(^3P_1)$ atoms collide with $(^1S_0)$ atoms, some of them fall into the $(^3P_0)$ metastable state. Thus, in the steady state two types of excited atom are present in the irradiated Hg vapour, so atoms of the two populations collide, and associate, whereby ionization occurs. Note that a molecular ion is formed in this reaction

$$Hg*(^3P_1) + Hg*(^3P_0) \rightarrow Hg_2^+ + e$$

because $V_i(Hg_2) = 9.6$ V which is balanced by $4.86 + 4.66 + 0.06 +$ K.E., where 0.06 V is the dissociation energy of Hg_2; remember $V_i(Hg) = 10.43$ V. This process is termed 'associative ionization', its cross-section is high ($\sim 5 \times 10^{-14}$ cm^2) and its rate is proportional to the (light intensity)^2, as expected. Zn, Cs and Cd seem to be similarly ionizable by resonance radiation, namely in two steps. We conclude that it is possible to ionize certain gases by light of quantum energy well below the ionization energy. Laser light of any wavelength can ionize or excite any molecule; associative excitation, Callear (1981).

The photo-excitation of molecular gases is a more complex problem. Assume visible monochromatic light is transmitted through molecular hydrogen. Does excitation of $(H_2)_{v=0}$ to higher vibrational states of the same electronic state take place? Since all diatomic homopolar gases have zero electric dipole moment they do not absorb (but scatter) radiant energy. However, when ultraviolet light of λ of the order 100 nm is used, excitation to a vibrational level of higher electronic states of H_2 (figure 2.11) takes place

$$H_2(^1\Sigma_g) + h\nu \rightarrow H_2^*(B^1\Sigma_u^+ \quad \text{or} \quad C^1\Pi_u \quad \text{or} \quad a^3\Sigma_g^+)$$

since radiant energy is actually absorbed. There are rules for selecting transitions that are possible; in general, no change in electron spin is permissible, except in the case of molecules (and atoms) whose spin is strongly coupled with the orbital motion of the electron, as in Hg.

Photo-ionization

By irradiating a gas with ultraviolet light of wavelength such that $hc/\lambda > eV_i$, (figure 2.6 and table 2.4) the gas can be ionized at a rate dependent on the photo-ionization cross-section q_i which is a function of λ. Figure 2.15, showing $q_i = f(\lambda)$

Table 2.4. Ionization potential V_i of molecules (in V).

H_2	N_2	O_2	C_2	F_2	Cl_2	Br_2	I_2	Hg_2	CH_4	C_2H_2	C_6H_6	CO	CO_2	NO	NO_2	NH_3
15.6	15.5	12.1	12	16.5	11.5	10.7	9.4	9.8	14.5	11.6	9.6	14	13.7	9.25	11	11.2

(For V_i of atoms see figure 2.6.)

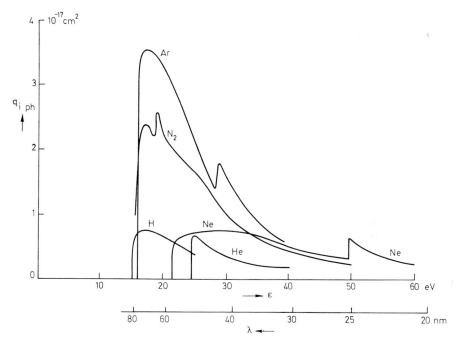

Figure 2.15. Photo-ionization cross-section $q_{i\,ph}$ of various gases, showing dependence on the quantum energy ε and the corresponding wavelength λ.
$\lambda(nm) = 1240/V_i(V)$.

for various gases, indicates the sudden rise of q_i at eV_i. Later, q_i decreases with rising energy ε (falling λ) interrupted by a series of discontinuous changes. These occur when the quantum ε has a value necessary to liberate an electron from the next closed shell. For example, a quantum ε of about 16 eV will ionize Ar by removing an electron from the outermost shell, whereas a 250 eV quantum can just extract an electron from the $(n = 2)$ L shell and a 3200 eV quantum can do likewise from the $(n = 1)$ K shell.

In the majority of cases, photo-ionization does not play the main role, for example in electric discharges, though it can be of great importance as a secondary source. However, it is the active process in ionization chambers for X-ray intensity measurements, in counters and other instruments.

2.7. Emission of charged and neutral particles from solids

Solids and liquids, whether conductors or insulators, are known to emit electrons, positive and negative ions and neutral excited particles, under certain physical conditions. These are: elevated temperatures, bombardment with beams of fast charged or neutral particles, encounters with slow-moving particles, irradiation with light, large electric fields, and so on. We shall consider here only some of the methods.

Thermionic emission

Electrons as well as positive ions are emitted from suitable hot cathode surfaces. The saturation electron current density (*in vacuo*) is given by Richardson's equation

$$j = AT^2 \exp\left(-e\phi/kT\right) \tag{2.7}$$

T being the absolute temperature in K, ϕ the work function in V and kT/e is the volt equivalent of T, $1\ V = 11\ 600\ K$. Values of ϕ are of the order $0.5\ V_i$ and are given in table 2.5.

It follows that the electron current intensity emitted from a plane cathode of pure W ($\phi = 4.5\ V$, $A = 7 \times 10^5$) at $T = 2500\ K$ is $j \sim 4 \times 10^3\ A\ m^{-2}$. The saturation electron current j does not reach the anode at low anode voltages because the distribution of initial energies allows only a certain fraction of the electrons to cross the potential minimum which is caused by the space charge in front of the emitter. In addition, in the presence of a gas a considerable fraction of electrons is back-scattered by elastic encounters within the first few mean free paths.

Equation 2.7 can be applied to ion emission when ϕ^+, the positive ion work function, is used instead of ϕ, together with another constant. In general $|\phi^+| > \phi$, say 6 V for W, and thus the positive ion currents are much lower, about $0.1\ A\ m^{-2}$ at 2800 K. Sodium and potassium salts give larger j^+ up to $100\ A\ m^{-2}$ at lower T. Negative ions are readily obtained from lithium and aluminium silicates.

Electron emission from cold surfaces by incident particles

When electrons of given energy ε strike a surface of a metal or insulator *in vacuo*, secondary electrons are emitted. The total number escaping in all directions per incident electron is δ, the secondary emission coefficient, which depends on ε and the substance (figure 2.16). All curves pass a maximum which is explained thus: at low ε the disturbance caused in the 'sea of electrons' by slow primary electrons is small, rises with ε and is confined to the upper surface layers. An electron of large ε penetrates deeper the larger ε is, but though the disturbance is more powerful only few of the accelerated deep electrons reach the surface and can escape.

Hence a maximum δ at medium ε results. For many substances, $\delta = 1$ occurs twice, at a low and a high value of ε, when for each incoming electron one

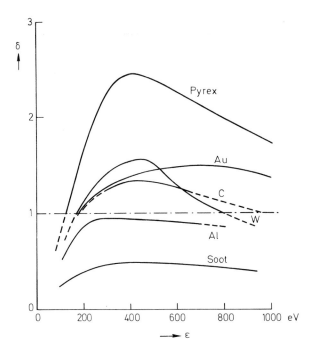

Figure 2.16. Secondary electron emission coefficient δ as a function of the energy ε of the incident electron for different substances.
δ is in secondary electrons per primary which strike the surface normally.

secondary leaves the surface; thus, an insulator struck by electrons charges up to a potential corresponding to the high value of ε for reasons of stability.

The more oblique the angle of incidence, the larger δ is ($\varepsilon =$ constant) because of the smaller depth of penetration of the electron. Using composite surfaces, such as oxide cathodes, instead of pure metals the values of δ can rise up to ten and more. The time interval between impact of the primary and emission of a secondary is less than 10^{-12} s.

Positive ions falling on a surface can also cause ejection of secondary electrons. For each secondary electron emitted, two electrons have to leave the solid, of which one neutralizes the ion, which rebounds as a neutral (occasionally excited) particle. In general, positive ions of low kinetic energy can release only a few electrons. Between 10 and 10^3 slow rare gas positive ions impinging on alkalis and ordinary metals, respectively, emit one secondary electron, i.e., the secondary electron emission coefficient $\gamma_i = 0.1-10^{-3}$. However, for fast ions of $\varepsilon \leqslant 10$ keV, γ_i rises up to three and more. The nature of the surface layer (impurity sites, metal oxide layer) has a large influence on γ_i, especially at low ε.

Metastable atoms and molecules at thermal energy can release electrons from solids with relatively high efficiency, depending on their surface structure. $\gamma_{met} \sim 0.1-0.5$ electrons per metastable are values quoted. The energy difference

Table 2.5. The work function ϕ (in V) of solids.

Li	Cs	Hg, Fe, Cu, W	Pd, Re, Nb	Zn, Ca, Mg, Si, U, Th, Hf	Cu$_2$O, CuO, Pt	WO$_3$	H$_2$O	Ba	BaO
~2·5	1·9	4·5	5	3·4–3·6	5·2–5·4	9·2	6	2·5	1·0

$V_{exc} - \phi$ is the maximum energy transferable to the secondary electrons. For metastable molecules, γ_{met} can be very low, as expected for systems with a large number of degrees of freedom.

Ordinary neutrals are also able to set free secondary electrons, but the emission coefficient is much smaller than that of the parent ion for slow particles and becomes equal for fast particles (potential and kinetic effect).

Electron emission from solids by irradiation

When monochromatic radiation falls on the surface of a solid of work function ϕ, photoelectrons are emitted when the quantum energy $h\nu > \phi$. It

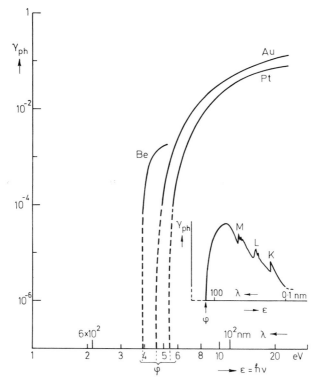

Figure 2.17. Photoelectric yield γ_{ph} of different solids as a function of the quantum energy ε. Ordinate in photoelectrons/quantum absorbed. Inset: $\gamma_{ph} = f(\varepsilon)$ for $\varepsilon \gg \varphi$.

follows that the longest wavelength λ_m for emission to occur is

$$\lambda_m < 1240/\phi \tag{2.8}$$

when λ_m is in nm and ϕ in V (see table 2.5). The photoelectric yield γ_{ph}, the number of electrons emitted per incident quantum, as a function of λ (or $h\nu$) is shown in figure 2.17 for some clean metal surfaces. First, γ_{ph} rises steeply for λ near the onset, at λ_m it passes a maximum at $h\nu \sim 10$–50 eV (~ 124–25 nm) and then decreases rapidly, interrupted by a series of discontinuities, namely the absorption edges in the X-ray region. This means that whereas with light of longer λ (low $h\nu$) the outer electrons in the conduction band of metals are extractable, with light of λ near $\gamma_{ph_{max}}$, electrons from lower levels in the band, and beyond $\gamma_{ph_{max}}$ electrons from the inner shells, are emitted. As to the energetics of photoemission, obviously thick metals reflect and absorb most of the incident quanta whereas thin foils are transparent. Absorption of light occurs by exciting the metal electrons, which dissipate the energy by interacting with 'phonons' (the quantum equivalent of the lattice vibrations) of energy of order 10^{-2} eV and less.

Positive ion emission from surfaces

A neutral atom of ionization potential V_i that strikes a metal surface of work function ϕ will shed one of its electrons on the metal when $\phi > V_i$ and then return

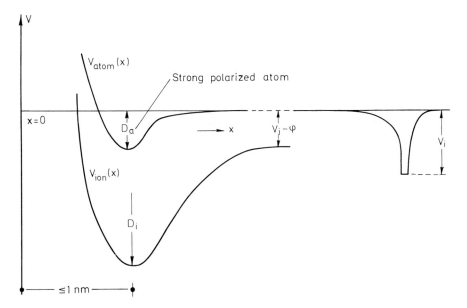

Figure 2.18. Adsorption potential $V(x)$ of a polarized atom approaching the wall $x=0$.
D_a = attractive potential energy minimum; $D_i = V_{im}(x)$ minimum. For Cs vapour striking a W wall; $V_i - \varphi < 0$. On the right: atom at large x (Dekker 1964, p. 228).

as a positive ion into its gas (figure 2.18). Alkali metals and high ϕ surfaces fulfil this condition and act as copious positive-ion sources. Technically, the 'cold' surface has to be kept at elevated T so as to avoid condensation of vapour which would reduce $(\phi - V_i)$. The number of positive ions per neutral leaving the surface is of the order exp $\{[\phi - V_i]/kT\}$. Thus, for $kT \sim 1200\ K \sim 0.1\ V$ and Cs on W with $\phi - V_i = 0.6\ V$, we find that for about 400 Cs$^+$ ions one Cs atom returns into the gas. On the other hand, at the same temperature, by directing a sodium beam $(V_i = 5.1\ V)$ on tungsten we would obtain exp $\{(4.5 - 5.1)/0.1\} = e^{-6}$ or 400 neutral Na atoms per Na$^+$, showing that small ion currents can be obtained even when $\phi < V_i$. If the tungsten electrode is heated to a higher temperature (say 0.2 eV or 2300 K) electron and ion emissions (ratio $= e^{-3}$, ~ 20 neutrals per ion) occur, so that the surface ionization source now produces a plasma beam.

Electron and ion emission from surfaces by strong electric fields

Large electric fields that develop on needle points, small projections or surface irregularities can be produced by applying over 10^4 V between the negative point and a far positive plate or grid. If spherical geometry and classical physics hold, the field E acting on an electron in the metal and causing emission would have to overcome image potential and work function ϕ of the order ϕ/r_0; with $\phi \sim 5$ V and $r_0 \sim 0.5$ nm, then $E \sim 10^{10}$ V m^{-1}. In fact, lower values of E are

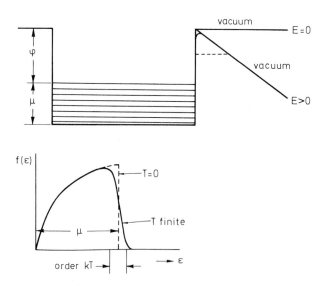

Figure 2.19. Top: Trough (box) model representing electron potential energies in metals with respect to vacuum.
Bottom: Fermi electron energy distribution $f(\varepsilon)$.
μ = Fermi energy; no electrons above μ at 0 K; E = electric field: top $----$ = electron tunnelling path; ϕ = work function.

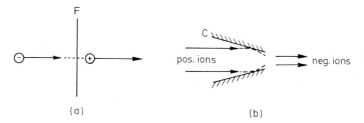

Figure 2.20. Double charge transfer.
Conversion of (*a*) a negative into a positive ion by passing the negative through a thin foil *F* which picks up 2 electrons (example: $H^- \rightarrow H^+$); (*b*) a positive into a negative ion by striking a solid surface obliquely (examples: H^+ (hitting Mg cone)$\rightarrow H^-$; Co^+ or Ar^+. 10–20 keV hitting C (15° cone)$\rightarrow C^-$, $\varepsilon \sim 200$ eV).

required and wave mechanics shows that the electron (de Broglie) wave can penetrate a potential wall by tunnelling through it with a probability which is larger the thinner the wall (figure 2.19). Theory leads to an expression for the electron current density of the form

$$j = c_1 E^2 \exp\{-c_2 \phi^{3/2}/E\} \tag{2.9}$$

where $c_1 = f(\mu, \phi)$ and c_2 is a constant. For $\phi = 4$ V, $\mu = 10$ V (Fermi limit) and $E = 3 \times 10^9$ V m^{-1}, we find $j \sim 10^5$ A m^{-2}.

Positive ions are also emitted from small points by fields about 10 times larger. A more copious ion source uses liquid metal which either covers a thin needle or fills a capillary tube to which several kV or more are applied. Ion currents of 10^{-4} A are so obtained from a single positive cusp of caesium, gallium, copper, mercury and other metals which is drawn out by the high field.

Negative ion emission from cold surfaces

Negative ions have been observed at very low pressure (surface ionization) when electronegative gases come into contact with tungsten. Another source of negative ions for nuclear accelerators uses a Cs^+ 20 keV ion beam of order mA which strikes, at a glancing angle, the inner surface of a hollow carbon cone (15° half angle) from which C^- ions are 'sputtered' and subsequently accelerated to about 10–40 kV. Beam currents of 20 μA of C^- are so obtained with an efficiency of 10^{-3} to 10^{-4} negative ions per incident positive ion. O^-, Cl^- and H^- beams can be produced in a similar way (figure 2.20). This process is analogous to the double-charge transfer process of fast positive ions in thin metal foils when the positive ion picks up two electrons from the foil and is thus converted into a negative ion.

CHAPTER 3

collisions, cross-sections and free paths

3.1. Introduction

Consider two neutral particles which move with moderate speeds along intersecting flight paths until colliding at the crossing point, where they can either be scattered in different directions or they can associate, dissociate or rearrange their atoms. If the particles were perfect elastic indivisible solid spheres of known mass moving *in vacuo*, linear and angular momentum (the latter because of the finite size) would be conserved. Therefore, velocities and directions after a collision or changes of the axes of the rotating spheres with respect to the initial directions could be determined. However, the results of a single collision between particles in liquids and in gases at moderate pressure, as regards momenta and directions, is often of less interest than the magnitudes averaged over a large number of collisions.

It is not yet possible to make visible the motion of single atoms (1 nm diameter); however, cigar-shaped gamboge particles have microscopic size (10^{-5} m length) and hence, by floating on a liquid, their translational and rotational motion can be easily observed with an ordinary microscope (figure 3.1(a)). J. Perrin measured before 1912 the mean distance $\bar{x}$ through which a test particle is displaced, and the mean angle $\bar{\theta}$ through which the rod-like particle turns, during a time interval t. He found that in a fluid both the mean square linear ($\bar{x}^2$) and the angular displacement ($\bar{\theta}^2$) vary with t. According to Einstein, not only is $\bar{x}^2 \propto t$ but $\bar{\theta}^2 \propto t$. The number of individual path lengths and angles are distributed according to Maxwell and Boltzmann. We note (a) that the constant in these proportionalities is the diffusion coefficient D, as, for example, in $\bar{x}^2 = 2Dt$; (b) that $\bar{x}^2$ and $\bar{\theta}^2$ result from the zigzag (random) motion which ceases when $T \rightarrow 0$; and (c) that no constant speed is associated with the displacement $(\bar{x}^2)^{1/2}$, since it would depend on time.

A treatment of the motion of colliding particles gives the distribution of velocities or energies for a gas in thermodynamic equilibrium (figure 3.2), but there is a different way to comprehend scattering.

In classical kinetic theory the gas molecules are taken to be hard elastic

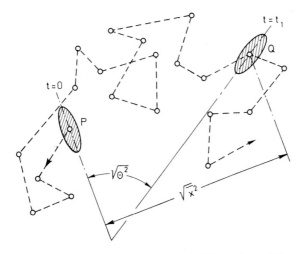

Figure 3.1. (*a*) Typical molecular one-dimensional path kinked by collisions (16) in a gas. illustrating the path length $l = PQ$ (~0·4 m), the mean displacement $(\bar{x}^2)^{1/2}$ (0·1 m) and the mean angular deflection $(\bar{\theta}^2)^{1/2}$ (52°, ~0·9 rad).

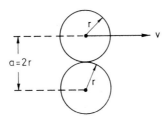

Figure 3.1. (*b*) A molecular collision.
r = molecular radius; q = collision cross-section ($q = a^2\pi$); v = relative velocity.

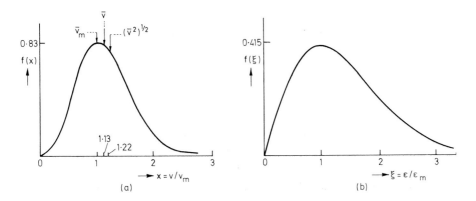

Figure 3.2. (*a*) Maxwellian velocity distribution.
(*b*) Normalized energy distribution.

spherical particles. Assume that a beam of these moves with an average speed v along a straight path through an assembly of like, stationary, uniformly distributed particles (see figure 3.2); whenever a beam particle touches a stationary one, a collision is counted. If z_0 particles (m^{-2} s^{-1}) enter the 'frozen' gas at $x=0$ and if z_x particles at x have escaped scattering between 0 and x, dz particles will collide between x and $x+dx$. This number will rise the larger z_x, N, the number density of stationary particles, and q, the cross-section of beam particles are. If a is the molecular diameter, here equal to the closest distance, then $q=a^2\pi$. Since d$z=z_x Nq$ dx, with $z_x=z_0$ at $x=0$, one obtains for the fraction of nonscattered particles

$$z_x/z_0 = \exp(-Nqx) = \exp(-x/\lambda) \tag{3.1}$$

where $\lambda = 1/Nq$.

In general, q is a function of the relative velocity between the particles and of the nature of scattering (elastic, inelastic). If, however, instead of the frozen gas target the molecules are in thermal motion, we find a shorter mean free path, namely

$$\lambda = (\sqrt{2}\,Nq)^{-1} \tag{3.2}$$

Since the probability P of a collision in dx at x is $P_x^{x+dx}=P_0^x P_{dx}$, P_0^x being the probability of no collision occurring from 0 to x, and P_{dx} the likelihood of a collision in any dx, one finds $P_x^{x+dx}=\exp(-x/\lambda)\,d(x/\lambda)$, $P_{dx}=1-\exp(-dx/\lambda)\approx(x/\lambda)$. Integrating P_x^{x+dx} between 0 and ∞ gives 1, i.e., certainty that all free paths lie within these limits. Thus, P_x^{x+dx} is the number distribution of free paths (figure 3.3) as confirmed experimentally by Born.

For a small atom such as He, an upper value of λ can be estimated by taking the Bohr radius $a_0=0.53\times10^{-10}$ m. With $q=4.2\times10^{-20}$ m^2 and $N=2.7\times10^{25}$ m^{-3} at standard pressure and temperature, we find $\lambda\sim6\times10^{-7}$ m,

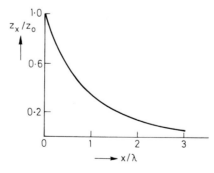

Figure 3.3. Distribution of free paths at room temperature. Attenuation of a beam of Ag atoms in air at low pressure.
(M. Born and E. Bormann, 1921.)
λ = mean free path.

instead of about 3×10^{-7} m, as observed. Thus, q is larger than the 'geometric' value of $q = 4a_0^2 \pi$ but the estimate gives the right order of magnitude.

In an ideal gas of density p the mean free path is

$$\lambda \propto 1/p \propto T/p \qquad (3.3)$$

but this is only approximately true. Although the dependence of λ on p holds over a very large range, say between 10^{-2} and 10^5 Torr, it becomes meaningless when wall collisions predominate and therefore usually fails when $p < 10^{-2}$ Torr because the vessel's size restricts λ, and is also inaccurate at high pressure when intrinsic molecular forces come into play. For $300 < T < 600$ K an equation of the form $\lambda(T) = \lambda_1/(1 + C/T)$, where λ is the corrected value of λ at 300 K and C, the 'Sutherland constant' holds; but for very high temperature, of the order 10^4 K, as in arc columns and plasma flames at and above 1 atm, a relation

$$\lambda(T) \simeq \text{constant } T^{5/4} \qquad (3.4)$$

holds. A curve of $\lambda(T)$ in N_2 at 1 atm corroborating equation 3.4 was derived from extrapolated viscosity measurements and calculations and is shown in figure 3.4; the dashed curve represents equation 3.3.

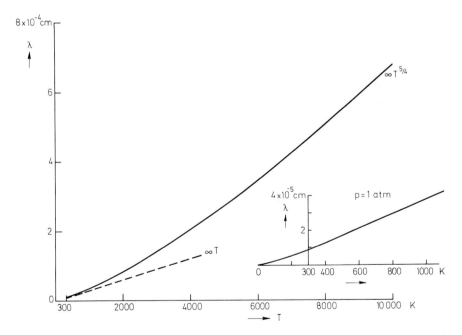

Figure 3.4. Mean free path λ of N_2 at 1 atm as a function of the gas temperature T in thermodynamic equilibrium.
(Derived from Amdur and Mason, 1958, *Physics of Fluids* **1**, 370, and Svehla, 1962, Nasa—TR.)
Inset: $\lambda = f(T)$ at lower T.

3.2. Free paths

The temperature dependence of the mean free path, or the decrease of q with rising temperature, may be thought to result from two effects: at low temperature, an increase in temperature favours attraction because the potential well (figure 3.5(a)) caused by the stray field of the nuclear charge raises the collision frequency v/λ and q. However, at high temperature repulsion by collisional compression and deformation of the outer electron cloud occurs, which increases with the particles' relative velocity and temperature. Repulsion is the dominant effect that reduces the apparent molecular size and hence q.

The values of λ for different elements (table 3.1) show satisfactory consistency as regards the periodic table: as the atomic number Z and the size of

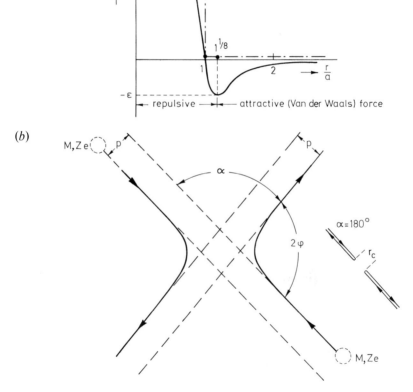

Figure 3.5. (a) Repulsive and attractive potential V_r dependent on the relative distance r/a between spheres.
$r/a = 1$ from equation 3.8 with $V_r = 0$. $- \cdot - \cdot - = V(r/a)$ for hard elastic spheres and zero attractive force $\mathrm{d}V/\mathrm{d}r$ for $r/a \geq 1$.
(b) Paths of two protons moving in opposite directions with the same velocity.

rare gas atoms rises, λ decreases. According to equation 2.1, the atomic radius $r \propto n^2/Z$ and thus the target area $q \propto (n^2/Z)^2$. Hence, $\lambda \propto Z^2/n^4$, which is seen to hold for He to Xe; here, Z varies from 2 to 54 and n from 1 to 5. Again, for organic molecules the larger their size the smaller λ becomes. In general, excited particles have a larger size and λ is smaller than for the parent particles in the ground (electronic or vibrational) state.

To picture molecules as billiard balls is often unsatisfactory, and for this reason the force concept has replaced the idea of the classical collision. In an elastic scattering collision the electron clouds of two approaching molecules repel each other (figure 2.13). On collision these clouds undergo a change in shape and interpenetrate slightly, thereby being 'in touch' for a very short but finite time. The 'contact or sticking time' rises with increasing interpenetration, which affects the mean free path. Since molecular forces decrease rapidly with distance from the nucleus, the path between two successive collisions is mainly straight, provided external fields of any kind are absent. However, at small distances the molecular field is strong and the path will be bent, its radius of curvature being smaller the closer the particles. Note that the difference between this and the 'solid sphere collision' is the suddenness of the change in direction; the kinked path of the solid sphere model is replaced by a smooth curved path. This idea also clarifies the meaning of collisions between pairs of neutrals, pairs of charged

Table 3.1. Molecular mean free path λ of atoms in their own gas at 760 Torr, 288 K and 1 Torr, 273 K.

Gas	$\lambda_{760}{}^{a,b}$ (m)	$\lambda_1{}^c$ (m)
He	$18 \cdot 6 \times 10^{-8}$	$13 \cdot 1 \times 10^{-5}$
Ne	$13 \cdot 2 \times 10^{-8}$	$9 \cdot 2 \times 10^{-5}$
Ar	$6 \cdot 7 \times 10^{-8}$	$4 \cdot 6 \times 10^{-5}$
Kr	$5 \cdot 1 \times 10^{-8}$	$3 \cdot 5 \times 10^{-5}$
Xe	$3 \cdot 8 \times 10^{-8}$	$2 \cdot 6 \times 10^{-5}$
H_2	$11 \cdot 8 \times 10^{-8}$	$8 \cdot 3 \times 10^{-5}$
N_2	$6 \cdot 3 \times 10^{-8}$	$4 \cdot 4 \times 10^{-5}$
O_2	$6 \cdot 8 \times 10^{-8}$	$4 \cdot 7 \times 10^{-5}$
H_2O	$4 \cdot 2 \times 10^{-8}$	—
CO_2	$4 \cdot 2 \times 10^{-8}$	$2 \cdot 9 \times 10^{-5}$
HCl	$4 \cdot 4 \times 10^{-8}$	—
NH_3	$4 \cdot 5 \times 10^{-8}$	—
CH_4	$5 \cdot 2 \times 10^{-8}$	—
C_2H_4	$3 \cdot 6 \times 10^{-8}$	—
C_2H_6	$3 \cdot 2 \times 10^{-8}$	—
Na	—	$4 \cdot 5 \times 10^{-5}$
Cd	—	$6 \cdot 9 \times 10^{-5}$
Hg	$\sim 3 \cdot 0 \times 10^{-8}$	$2 \cdot 2 \times 10^{-5}$

[a] At 1 atm, λ_{760} is of the order of 100 molecular diameters, which justifies the notion of 'molecular chaos'.
[b] From McDaniel, 1964.
[c] From Schulz, 1968.

molecules or encounters between neutral and charged particles, though in the latter case 'charge exchange' (see p. 50) may occur.

3.3. Collisions between two ions

A formally simple case is the collision between two protons (H^+) moving in opposite directions with the same velocity with respect to the laboratory system. Again, let the particles move in a field- and otherwise particle-free space (figure 3.5(*b*)). If one and two were neutral point masses, they would pass each other at the distance p_c. Because of their positive charges their velocity in the initial direction and their kinetic energy is reduced, while the electric field and potential between them first increases. From a drawing of the field lines for different positions, the density of lines and the mutual repulsion will be found to rise, reaching a maximum when the particles come closest. (The reader is advised to sketch the field.) Since no energy loss is envisaged here, the initial and final velocity of each particle must be the same but their direction has changed by an angle α. The particles have undergone a single elastic scattering; the theory was first enunciated by Rutherford.

Instead of deriving the path and the number of particles scattered into different angles, we shall deduce a relation (for equal particles of charge Ze) between the initial energy E_0, and the area of the scattering target, the scattering cross-section q. Consider a collision between a pair of particles initially moving in opposite directions, along two parallel lines separated by p, the impact parameter. If $p = 0$, reflection occurs, i.e., scattering through $\alpha = 180°$; at the instant of closest approach (p_c) the velocities are zero and the potential energy of each particle of mass M is equal to E_0. For a head-on collision, energy conservation in a coulomb field gives

$$E_0 = \tfrac{1}{2}Mv_0^2 \simeq \tfrac{1}{2}(Ze)^2/p_c \tag{3.5}$$

and the distance of closest approach is

$$p_c \simeq (Ze)^2/Mv_0^2 \propto 1/E_0 \tag{3.6}$$

For $\alpha = 180 - 2\phi$ one finds $\tan \phi = p_c Mv_0^2/(Ze)^2$ (see figure 3.5(*b*)). From equation 3.6 it follows that the larger E_0 and the lower the atomic number Z, the nearer the particles approach each other. The scattering cross-section for $\alpha = 90°$ ($\phi = 45°$ and $\tan \phi = 1$), which we obtain from above, is

$$q_{90} = p_c^2 \pi = \pi(Ze)^4/M^2v_0^4 \propto 1/E_0^2 \tag{3.7}$$

This depends sensitively on E_0 and on Z. Moreover, it can be shown that $(v_0^4 \sin^4 \alpha/2)^{-1}$ determines the intensity I of angular scattering, i.e., the fraction of particles scattered into α, and therefore relatively few fast particles are scattered into large angles. The dependence of q on E_0 or v_0 can be easily understood: as v

rises, the interaction time, during which the particles are sufficiently close to feel the electric force, becomes shorter and both the total momentum exchanged $\left(\int f \, dt = \int m \, dv\right)$ and q decrease. This, however, holds only when α and v_0 are large; 90° scattering will then occur seldom since the particles may pass each other at large distances. Yet from equation 3.7 it follows that when $v_0 \to \infty$, p and $q \to \infty$, which is unreasonable. This infinity problem arises from the use of the infinite range of the Coulomb potential $V \propto 1/r$ which we applied in equation 3.5 to two interacting charges. The physical argument is faulty because the necessary corresponding negative charges have been placed at infinity (the universe is thought to have zero net charge). When that error is corrected, for example by using a 'screened Coulomb potential' of the form $V \propto \exp(-cr)/r$, c being a constant, the infinity in q disappears.

3.4. Collisions between two neutral particles

When two neutral atoms are about to collide, the spatial electric potentials V_r due to their strong stray fields cannot, even approximately, be expressed either as Coulomb or screened potential (figure 3.5(a)). A rather convenient way is to split up the interatomic potential into an attractive and a repulsive term (which is of physical significance) and use their difference. For simple molecules it reads, for example

$$V(r) = 4[(a/r)^{12} - (a/r)^6] \tag{3.8}$$

where a is the atomic diameter and r the interatomic distance measured between the two centres. Figure 3.5(a) shows that this function has a zero at $r = a$, that when $r < a$, V_r increases very steeply with a slope $dV/dr < 0$ indicating a repulsive force, that the minimum ($dV/dr = 0$) lies at $r/a = 2^{1/6} = 1 \cdot 125$ and that $V_{r\,min} = -\varepsilon$ (which is easily verified), ε being the maximum attractive potential at the equilibrium position ($1 \cdot 25$). The dash-dotted curve represents V_r for ideal hard elastic spheres in contact ($r = a$) and zero van der Waals forces. The physical basis of V_r has been explained earlier (p. 37).

Treatments of the kind presented here also fail when Heisenberg's uncertainty relation has to be invoked, i.e., when wave rather than particle concepts must be applied to obtain a physical insight into a process. Since in the case of figure 3.5 the product of the uncertainties of linear momentum $\Delta p = \Delta(Mv)$ and position Δx is

$$\Delta p \Delta x \sim \hbar \tag{3.9}$$

then with $\hbar = 10^{-34}$ J s and $M \simeq 1 \cdot 7 \times 10^{-24}$ g, a head-on proton–proton collision corresponding to equation 3.6 gives $\Delta x = r_e \sim 4$ nm, or approximately 40 atomic diameters when ε is more than 10^6 eV. Thus, when E_0 is of order $\Delta \varepsilon \sim 10^6$ eV or less, diffraction (or wave) effects, i.e., oscillatory variations of q with ε (and of intensity I with α), can be expected; the magnitude of the amplitudes is not easily predictable. Experimentally, this effect is not observable for proton–proton collisions, but for electron–atom and electron–molecule encounters it is often very strongly pronounced. Since the electron mass is more than 10^3 times smaller than that of the proton and the atomic size is approximately 10^5 times larger, $\Delta\varepsilon$ is of order 10 eV.

Experimental evidence of what is called the Ramsauer–Townsend effect is given in figure 3.6, showing the total elastic electron scattering cross-sections for some of the rare gases. The trend is for increasing scattering cross-sections with increasing atomic number, the cross-section of Hg being very high, for example. It was first detected in Ar, where electrons of energy of about 1 eV suffer a minute 'scatter'; in spite of an 'atomic target' of 2×10^{-19} cm^2, such slow electrons have an m.f.p. of about 2/3 cm at 1 Torr gas pressure. This means that diffraction of the 'electron waves' around the 'opaque' atomic obstacles takes place; in this way scattering into all angles behind the target smoothes out the diffraction pattern, thus masking the presence of a target. (This does not apply to the angular (α) dependence of the differential cross sections, α being the angle into which scattering occurs.)

Another way to express the condition under which classical thinking has to be replaced by wave-mechanical thinking is to compare the particle wavelength λ_m with the 'size' a of the target. Equivalent to equation 3.9 we find, as in optical systems

$$\lambda_m > a \qquad\qquad (3.10)$$

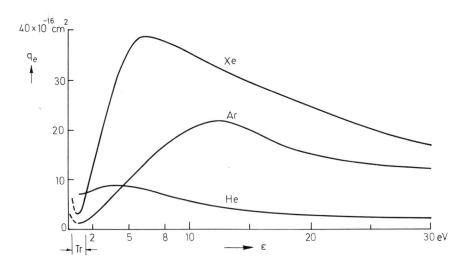

Figure 3.6. Ramsauer–Townsend effect = momentum transfer cross-section for electron-rare gas atom collisions $q_e = \mathrm{f}(\varepsilon)$.
ε is the electron energy in eV. Tr = transparency range or 'window' defined as $q_e \simeq 0$.

Since $\varepsilon = h\nu = hc/\lambda$ and $\varepsilon/c = p$, p being the quantum's linear momentum. $\lambda_m = h/p_m$ and $\varepsilon_m = p_m^2/2m$ $(p = mv)$. Thus, $\lambda_m = h/(2m\varepsilon_m)^{1/2}$. For an electron hitting atoms, λ_m in nm $= (1500/\varepsilon_m)^{1/2}$, for ε_m in V, so that for $\varepsilon_m = 10$ V, $\lambda_m \sim 0.4$ nm. Hence, equation 3.9 or 3.10 teaches that when the atomic diameter a is less than 0.4 nm, a 10 V electron will suffer diffraction, i.e., classical mechanics fails; experiments (figure 3.6) have confirmed this conclusion.

So far we have confined ourselves to elastic scatttering, a process associated with a partial exchange of energy between a fast and a slow particle whereby the fraction of the energy transferred is about twice the ratio of their masses. If the slow particle is heavy and the fast one light, such as in a collision between a molecule of mass M at thermal speed, and a fast electron of mass m, then by balancing linear momentum and energy (which for inelastic collisions must include a 'reaction', or potential, energy term Δ) the fraction f of the energy of the fast particle, transferred and converted in a head-on collision into potential energy of the molecule, is

$$f = \Delta/\varepsilon_1 = M/(M + m) \qquad (3.11)$$

showing that when m is much less than M nearly the full initial electron energy ε_1 can be used in performing a 'reaction' such as excitation, ionizaton or dissociation of the molecule. If there is an excess of energy, the electron will share it with the molecule, i.e., the former will acquire practically the whole excess energy in the form of kinetic energy.

The relation in equation 3.11 can also be applied to collisions between equal masses. In this case, $m = M$, and $\Delta/\varepsilon_1 = 0.5$, and thus to perform any of the 'reactions' envisaged, $2\varepsilon_1$ is the minimum initial energy required. This has been confirmed experimentally: ionization of the He atom by an electron arises when its energy exceeds 24.6 eV; yet the kinetic energy of an He atom must exceed 49.2 eV to ionize another He atom at rest ($He_{fast} + He \rightarrow He^+ + e + He$).

It is necessary to emphasize that if the energy available for a reaction is below a minimum or onset energy, the reaction occurs with zero probability and has zero cross-section. Thus, ionization of He ($V_i = 24.6$ V) does not occur when an electron of that critical energy hits an atom (which may become excited). However, an electron of energy ε greater than 24.6 eV has a finite chance to force the atom to expel an electron, thereby creating a singly charged positive ion and another electron.

Since we started above with a discussion of general reactions, we might as well write the ionization by electron collisions in the form of a chemical relation indicating the balance of particle mass and charge

$$e_{fast} + He \rightarrow He^+ + e_{slow} + e_{slow}$$

One of the slow electrons is the slowed-down 'primary' and the other the 'secondary', both usually having different energies, the energy sum being $\varepsilon_1 - eV_i$.

We shall now enquire about the value of the probability of ionization P_i by electron impact. Since P_i is the ratio of 'lucky' to 'total' number of events (collision cross-sections q_i, q_t) then

$$P_i = q_i/q_t$$

Its value depends on the electron energy ε and hardly exceeds approximately 40 per cent at about 100 eV. Since, by definition, $q_i = 0$ for $\varepsilon \leq \varepsilon_i$, and as any q_i is known to decrease with ε when ε is large, it follows that q_i and P_i rises from maxima at finite values of ε. In figure 3.7 the ionization cross-sections q_i as a function of the electron energy ε are shown for He and H_2. For He, q_i rises from zero at $eV_i = 24\cdot6$ eV to a value of $q_{i\,max} \simeq 0\cdot36 \times 10^{-20}$ m^2 at $\varepsilon = 100$ V and then decreases slowly. At the maximum, $P_i = 0\cdot14$, meaning that on the average every 7th collision of a 100 eV electron ionizes an atom. Similar curves are found for other gases (except alkalis). A more detailed investigation shows that electrons of $\varepsilon > 80$ V can doubly ionize He

$$e + He \rightarrow He^{2+} + 3e$$

and similar multiple ionizations occur in other gases. However, whereas the total ionization curve will not be much different from figure 3.7, because it entails a relatively small correction, the contribution is considerable, for example in argon.

When electrons of energy ε are shot into He and their losses due to inelastic collisions are measured, it turns out that at $\varepsilon > 19\cdot8$ eV, that is $4\cdot7$ eV below the

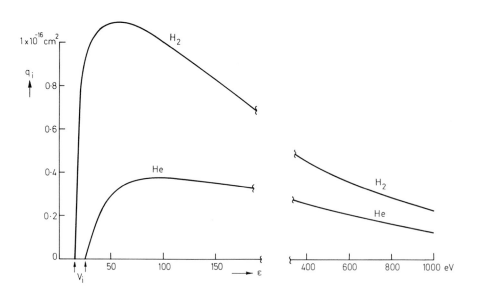

Figure 3.7. Ionization cross-section q_i by electron collisions as a function of the electron energy ε for H_2 and He.
$q_{i\,max}$ and ε_{max} for $H_2 = 1\cdot1 \times 10^{-16}$ cm^2 at 60 eV; for He $= 0\cdot37 \times 10^{-16}$ cm^2 at 90 eV.

Table 3.2. Free life of some metastable atoms and molecules of energy ε.

Species	State	Life (s)	ε (eV)
He	2^3S_1	$\sim 10^{-4}$	19·8
	2^1S_0	$\sim 10^{-2}$	—
Ne	$^3P_2$	—	16·7
	$^3P_0$	—	16·6
Ar	$^3P_2$	$3\cdot5 \times 10^{-3}$	11·7
	$^3P_0$	$3\cdot5 \times 10^{-3}$	11·5
Hg	$^3P_2$	>1	5·4
	$^3P_0$	1·6–2·5	4·7
H	$2^2S_{1/2}$	0·12	10·2
O	^1S	0·45–0·8	4·2
	^1D	150–200	2·0
N	^2P	—	3·6
	^2D	—	2·3
O_2	$^1\Delta g$	—	0·98
	$^1\Sigma g$	—	1·6
	$A^3\Sigma_u^+$	~ 1	4·5
N_2	$A^3\Sigma_u^+$	~ 12	6·2

ionization energy, losses set in which are not accompanied by emission of radiation or by delivery of new electrons. It transpires that excited He atoms, He*, in long-lived states are produced. Such metastable atoms lose their energy by either hitting the wall of the vessel or, on rare occasions by spontaneously

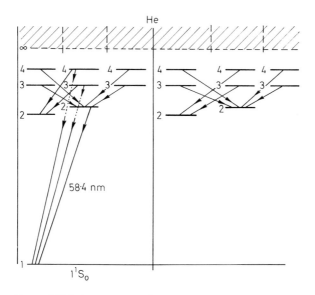

Figure 3.8. Transitions of He* down to $n \geq 1$ (emission between the visible and far u.v. singlet system) and down to $n \geq 2$ (emission between the near u.v. and visible + triplet system) such as $3^3P - 2^3S$ (388·9 nm) and $3^3D - 2^3P$ (587·6 nm).
Spin–orbit coupling negligible. Singlet–triplet transitions only for very large n.

emitting 50 nm photons, i.e., vacuum-ultraviolet radiation, or by colliding with 'impurity molecules' which abstract and convert this energy in various ways. The mean 'free' life time of the two lowest metastable states, 2^3S_1 at 19·7 V and 2^1S_0 at 20·6 V, is of the order 10^{-3}–10^{-2} s (table 3.2). Because of the long time which elapses until this potential energy is spontaneously released (like that stored in an expanded spring), the population of metastable He atoms can often become relatively large so that perhaps one in 10^4 atoms of a gas may be a metastable. Rare and other gas atoms (Ne, O) and metal vapour atoms (Hg, Cd) are known to form metastables as well as molecules (N_2, O_2), some with a free life of up to many minutes. However, the majority of high atomic energy states are those which remain excited for only a short time, 10^{-8}–10^{-11} s, after which they emit a light quantum while returning to the ground state.

A few principal radiating and metastable states are shown in the He level diagram figure 3.8. The states fall into two groups according to whether pairs of electron spins are in the same or opposite directions (triplet and singlet states have parallel and anti-parallel spins respectively). Since the spins of the two electrons of He in the ground state (singlet) are anti-parallel electron collision excitation into the metastable (triplet) state requires an electron exchange. Triplet is the name given to groups of three emission lines whose transitions start on three and end on

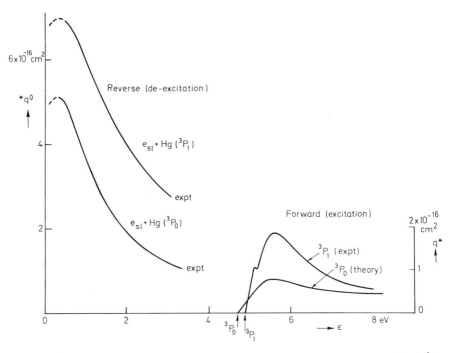

Figure 3.9. Superelastic (second kind, reverse, de-excitation) electron collision cross-section $*q^0$ in Hg $=f(\varepsilon)$ and electron excitation cross-section $q^*=f(\varepsilon)$ to the resonance state $^3P_1$. sl = slow.

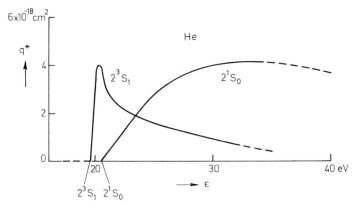

Figure 3.10. Electron excitation cross-section $q^*=f(\varepsilon)$ in He to $2\,^1S_0$ and $2\,^3S_0$ level. In Ne($^3P_{0,2}$, $16\cdot7$ eV) $q^*_{met}=1\cdot6\times10^{-18}$ cm^2 (max.) at $\varepsilon=20$ eV.

one energy level or vice-versa. Some of the levels of figure 3.8, the metastables, are not connected by lines with the ground state, indicating that these transitions are 'spectroscopically forbidden', i.e., not associated with emission by radiating dipoles. This means that the atom does not absorb or emit radiation or energy corresponding to the vertical distance between the upper level and that of the ground state. However, excitation and superelastic (de-excitation) electron collisions take place with finite likelihood; cross-sections for Hg are shown in figure 3.9 and data for other atoms are now available.

Electron excitation cross-sections have a common feature. Singlet to triplet excitation cross-sections q^* by electron collisions rise fast with the electron energy ε, their maxima being a few eV above the onset energy, whereas in singlet–singlet transitions q^* rises slowly with ε and thus q_{max} lies far above the onset energy (figure 3.10). These facts are important when a numerical estimate of the intensity

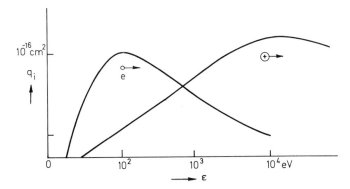

Figure 3.11. Comparing ionization cross-sections $q_i=f(\varepsilon)$ of electrons with those of ions in their parent gas.

of emission lines or the population density of excited and metastable atoms is made.

Atoms can be excited by collisions with fast atoms or ions but the energies at which $q = q_{max}$ are in general at least two orders of magnitude higher (figure 3.11). Another way is photo-excitation, i.e., using radiation from light sources which is absorbed by the gas atoms. However, this method is only effective for atoms with large absorption cross-sections (resonance radiation). The corresponding resonance lines result from transitions to the ground state. Well known intense lines are the Hg ($^3P_1-^1S_0$), 253·7 nm and the Na (doublet) resonance lines ($^2P_{1/2, 3/2}-^2S_{1/2}$), 589·0 and 589·6 nm where the lowest excited states 'combine' with the ground state. These resonance energies are given in table 2.3 (p. 23).

We now turn to the cross-sections associated with molecules. Here the situation is more complex. We shall deal first with the H_2 molecule as described in Chapter 2 and treat it as a dumb-bell rotating about two axes. Since rotational energies at 300 K are of the order 1–10 meV per molecule, the principal source of electron excitation to higher rotational levels is free electrons interacting with the permanent quadrupole moment of the bound electrons of H_2. If J is the rotational quantum number, the rotational energy of a single homonuclear molecule is approximately

$$E_{rot} = \text{constant} \times J(J + 1) = \{\hbar^2/2I\}J(J + 1) \qquad (3.12)$$

I being its moment of inertia. Thus, for $J = 3$ with $\hbar = 10^{-34}$ J s and $I = 10^{-44}$ g m², $E_{rot} \sim 6 \times 10^{-21}$ J $\sim 4 \times 10^{-2}$ eV. We note that the emission by a rotational transition can again only be due to the quadrupole moment of H_2, and selection rules demand that J changes by ± 2 (not by ± 1 as for dipoles). Figure 3.12 shows the dependence of q_{rot} on the electron energy ε for the transitions $J = 0 \to 2$ and $J = 1 \to 3$, starting at E_{rot} of about 5 and 8×10^{-2} eV respectively; q_{rot} rises steeply with ε and reaches saturation. For $E_{rot} > 0.8$ eV, $q_{rot} < q_{vib}$ and at still larger E, $q_{rot} \ll q_{vib}$. For N_2 the values of q_{rot} are of the same order as for H_2, but the dependence on E differs slightly. For gases with a permanent dipole, such as

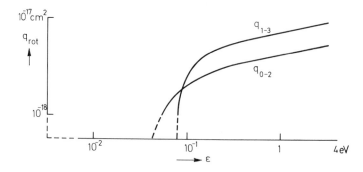

Figure 3.12. Rotational electron collision cross-section as a function of the electron energy ε for transitions $J = 0 \to 2$ and $J = 1 \to 3$ in H_2.

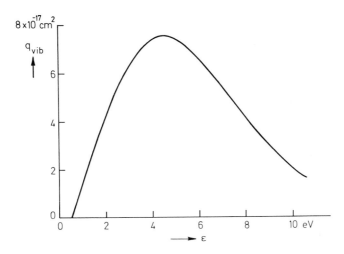

Figure 3.13. Vibrational electron collision cross-section $\Sigma_{v=0}^{v=\infty} \, q_{vib} = f(\varepsilon)$ in H_2.
For the potential energy diagram of H_2 ground state see figure 2.11.

H_2O, NH_3 and CO, q_{rot} reaches maxima up to two orders higher than those for N_2 or H_2.

The vibrational excitation of H_2 molecules by electron collisions from $v=0$ to $v>1$ requires electrons of $E>0.52$ eV. The vibrational cross-section $q_{vib} = f(\varepsilon)$ are very much larger than the rotational ones. The curve of figure 3.13 represents the sum of all individual vibrational transitions from $v=0$ to $v\to\infty$. The maximum of $q_{vib} \sim 8 \times 10^{-21}$ m^2 lies at $E\sim 4$ eV. It should be noted that neither the absolute values given nor the mechanism of vibrational excitation by electrons are certain.

The situation is different in N_2. Because of the larger nuclear mass the vibrational period T_v is only approximately 10^{-14} s. However, a slow electron incident upon N_2 forms an unstable N_2^- ion (see Chapter 2). The life of N_2^- is much longer than T_v. Finally, by temporary attachment, the electron with 2 eV energy leaves the molecule in a higher vibrational state. The unusually large cross-section of this process, first observed before 1930 ($\sim 3 \times 10^{-19}$ m^2), is notable and the energy loss in this 'resonance' collision is also considerable. Outside the range $E\sim 2$–3 eV, q_{vib} of N_2 is lower than that of H_2.

Electrons of $E>8.8$ eV which collide with H_2 give rise to an electronic transition from the ground state $^1\Sigma_g$ to the repulsive $^3\Sigma_u^+$ state (figure 3.14), i.e., to dissociation. The excitation path in the potential energy diagram is vertical, agreeing with the Franck–Condon rule, and gives the experimentally confirmed onset energy for two bound H atoms to separate with kinetic energy of 2.15 eV each. Table 3.3 contains comparative values for some molecules. The dissociation cross-section of H_2, $q_{diss} = f(\varepsilon)$, is shown in figure 3.14. It rises steeply to a maximum at approximately 16 eV and decreases rapidly, as expected of a singlet–triplet (spin-changing) transition in which a bound electron with an

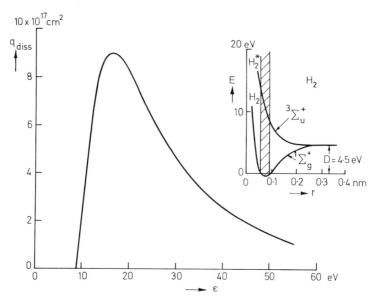

Figure 3.14. Electron collision dissociation cross-section $q_{diss} = f(\varepsilon)$ for H_2. Onset at 8.8 eV.

oriented spin is exchanged against incoming free electrons whose spin is not pre-orientated.

Ionization of homopolar molecules by fast electrons is of major interest. For example, for H_2 the onset (ionization) potential is given by the vertical distance between $(1^1\Sigma_g)_{v=0}$ and $^2\Sigma_g$ in figure 2.11.

The next process under discussion is of the type

$$\text{molecule} + e_{fast} \rightarrow \text{molecular ion}^+ + 2e$$

Figure 3.7 shows the electron ionization cross-section q_i as a function of the electron energy ε for H_2 whose character is the same for molecular gases. The shape of these curves and their maxima can be explained as follows: when ε is large an increase in ε will make the interaction time t_{in} between the fast primary electron and the slow orbital electron (t_{in} is of the order a/v_e, the time required to

Table 3.3. Thermal dissociation and electron collision dissociation energies of diatomic molecules.

Molecule	Thermal (V)	Electron (V)
H_2	4·5	8·8
O_2	5·1	~7
N_2	9·8	24·3
NO	6·5	>10
Cl_2	2·5	~3·7

cross the atom) gradually shorter and hence the exchanged linear momentum normal to the electron orbit becomes smaller. Thus, q_i decreases as ε rises. At low ε, not too far above onset energy, the velocity of the primary electron is of the same order as the orbital velocity of the atomic electron. Because of the mutual electron repulsion the primary electron which approaches the atom is always gradually deflected from one of the bound electrons. Therefore, when ε is increased, the bent path of the primary electron is straighter, a larger fraction of the momentum is transferred, and q_i rises as ε increases. The maximum in q_i results from joining the curve of the low ε region with the curve of the high ε region. A similar argument applies to all the other types of ionization and excitation cross-sections.

Another fairly frequent event is the ionization by electron collisions which is accompanied by dissociation. The general process referred to reads

$$\text{molecule} + e_{fast} \rightarrow \text{atomic ion}^+ + \text{atom} + 2e$$

For example, the dissociative ionization of molecular hydrogen is

$$H_2 + e_{fast} \rightarrow H^+ + H + 2e$$

Figure 3.15 shows $q_{i\,diss} = f(\varepsilon)$ for several diatomic molecules. Note that the maxima are at about the same values of ε as those in figures 3.7 and 3.16.

Figure 3.16 presents an overall picture of the various electron collision cross-sections in H_2 and draws attention to the contribution to the total cross-section. Clearly, the importance of elastic and inelastic processes changes because of their prominence in different ranges of electron energy, ε. It is obvious that in a

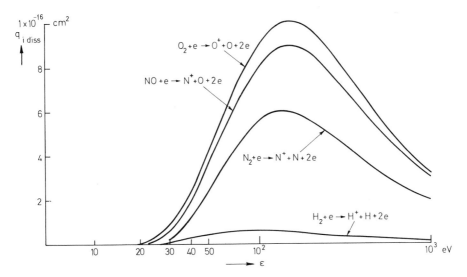

Figure 3.15. Dissociative ionization cross-section $q_{i\,diss}$ by electron collisions dependent on the electron energy ε for O_2, NO, N_2 and H_2.

molecular gas, strictly speaking, inelastic collisions occur even at very low ε (see figure 3.12), though these represent only a tiny fraction of the total cross-section q_t. However, above 1 eV the vibrational and excitation cross-sections, and above approximately 20 eV the ionization cross-section, become more prominent, until finally, for electron energies above 300 eV the ionization cross-section gradually approaches the (extrapolated) total cross-section. This means that at low ε elastic and at large ε inelastic collisions are of major importance. Similar graphs to that for H_2 (figure 3.16) have been produced for other molecular and atomic gases such as Hg (figure 3.17), confirming the general trend presented.

A collision process of a special kind of electron capture is the 'charge transfer' or 'charge exchange' collision. When a positive atomic ion moves through its own gas and collides with a neutral atom then, for low ion energies, the temporary psuedo-molecule which forms and persists during 'contact' suffers a rearrangement of charge. As a result of this the fast incoming positive ion becomes neutralized by extracting an electron from the atom but continues to move in the same direction without changing its kinetic energy. The originally neutral atom which had thermal speed and has lost an electron becomes a positive ion again without a change in speed. In the cases considered (He^+ in He, H_2^+ in H_2) no change in total energy and momentum occurs. The collision described is termed symmetric, or resonant, and the charge transfer cross-section is large (of the order of thermal elastic molecular cross-section, say $q_{+0} \times 10^{-19}$ m²). In

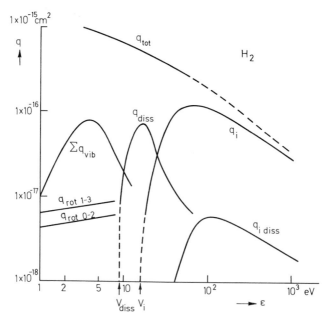

Figure 3.16. Comparing the total (momentum transfer) cross-sections of electrons $q_{tot} = f(\varepsilon)$ with q_{rot}, q_{vib}, q_{diss}, q_i and $q_{i\,diss}$ in H_2.

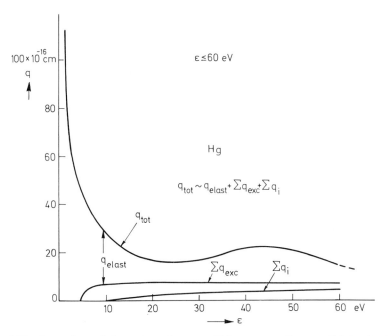

Figure 3.17. Comparison of the various electron collision cross-sections $q = f(\varepsilon)$ in atomic Hg vapour.

unsymmetric cases the ion elastic cross-section often exceeds q_{+0}. Ions can also undergo double charge transfer (two electron exchange).

The charge transfer cross-section $q_{+0} = f(\varepsilon)$ for He^+ ions in He is about 1×10^{-15} cm^2 at 100 eV. As the ion energy ε^+ increases, q_{+0} decreases slowly, reaching about 3×10^{-16} cm^2 at 10 keV; at larger energies additional processes are observed. However, the unsymmetric charge transfer cross-sections, such as for He_2^+ ions in He, are very much smaller, because the exchange of an electron is associated with a change in total energy during collision (figure 3.17). It would therefore be wrong to conclude that q_{+0} of He_2^+ in He, because of the larger size of He_2^+, is greater than that of He^+ in He. What matters in this non-classical charge exchange process is the fact that in the latter case of 'energy resonance' the total energy in the system is conserved, but this is not so when a molecular ion moves through its atomic gas or atomic ions through molecules. Just as for slow electrons in gases, the encounter of positive ions with molecules of their own gas cannot be described by a single classical collision process, but is the sum of classical scattering and charge exchange. This is seen in figure 3.18 for Ar^+ ions in Ar. It is a reminder that even in symmetric charge-transfer cases the elastic cross-sections cannot always be neglected.

The charge transfer process is often accompanied by excitation and dissociation. An atomic fast ion interacting with a slow gas atom A_s yields

$$A_f^+ + A_s \rightarrow A_f^* + A_s^+$$

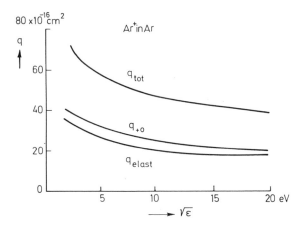

Figure 3.18. Comparison of the total (q_{tot}), charge transfer (q_{+0}) and elastic (q_{elast}) cross-sections of Ar$^+$ in Ar as a function of the ion energy ε^+.

and a fast molecular ion A_{2f}^+ can produce

$$A_{2f}^+ + A_s \rightarrow 2A_f + A_s^+$$

The double charge transfer is the exchange of two electrons, an example being

$$A_f^{++} + A_s \rightarrow A_f + A_s^{++}$$

The cross-section in this case has been found to be smaller than in single charge transfer, as expected.

It is of some interest to discuss a collision process in which large ions are formed by polarization attachment. An electron, or an ion of either polarity, which approaches a molecule of zero permanent electric dipole moment m is attracted and, under favourable circumstances, can become attached to the molecule when the force due to its polarizability α is strong enough. The capture of the charge occurs at the critical distance r_c (figure 3.19) for which the attractive potential energy P of the molecule is equal to the kinetic energy K of the ion. It can be shown (see Chapter 4) that the induced dipole field E which derives from the end-on position of dipole charges and the inverse square law is $E = 2\mu/r^3 = \alpha 2e/r^5$ and thus with $P = e \int E \, dr = K$ we find for $r = r_c$

$$\frac{\alpha e^2}{r_c^4} = \frac{Mv^2}{2} \tag{3.13}$$

The attachment cross-section is, therefore, neglecting a factor of the order one,

$$q_{att} \simeq r_c^2 \pi = \left(\frac{\alpha}{M} \right)^{1/2} \frac{e\pi}{v} \tag{3.14}$$

For thermal speeds v of about 10^3 m s^{-1} and molecules of mass about ten times that of H with $\alpha \sim 10^{-29}$ m^3 (see Chapter 2), q_{att} is of the order 10^{-18} m^2. We

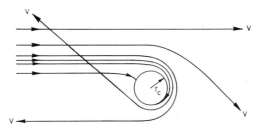

Figure 3.19. Trajectories of a charge (e, M) of velocity v for different impact parameters r. The moving ion interacts with a neutral stationary molecule of zero permanent dipole moment; for $r = r_c$, the critical impact parameter, the (open) trajectory goes over into a closed orbit and ion attachment occurs.

conclude that, compared with other cross-sections quoted earlier, q_{att} is of considerable magnitude. The reverse process—detachment—can be caused by collisions with quanta, electrons and neutral or charged particles.

3.5. Evaporation into vacuum and gas

By raising the temperature, or by other means, a solid or liquid surface bound by vacuum is compelled to emit gas (vapour) and the ensuing flow of particles will occur at a rate $\Gamma_{vac} = N\bar{v}/4$ where N is the number density (in m^{-3}) of surface particles and $\bar{v}$ the appropriate mean velocity (in $m\ s^{-1}$). Once emitted,

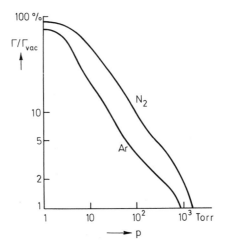

Figure 3.20. Net rate of evaporation (back-diffusion, back-scattering) of W atoms (in $g\ cm^{-2}\ s^{-1}$) emitted from a hot wire of diameter $\phi = 75\ \mu m$ at 2400 K, *in vacuo* (Γ_{vac}), and in Ar and N_2 (Γ) dependent on the gas pressure p.

no particle is due to return to the surface. However, when the bounding medium is a gas the emitted particles are back-scattered and the net flow rate is reduced to Γ.

Few observations on back-diffusion are recorded. Figure 3.20 shows $\Gamma/\Gamma_{vac} = f(p)$, the relative evaporation rate of a 75 μm thick W wire at 2400 K in Ar and N_2 at pressure p. A pressure in the range 3–8 Torr makes $\Gamma/\Gamma_{vac} = 0.5$. If an evaporated W atom should strike a wire of diameter d it must be back-scattered to condense on it. Thus its mean free path λ_W must be of the order d or less. Taking λ_W in $N_2 \sim \lambda_{Hg} \sim 2 \times 10^{-5}$ m at 1 Torr, 273 K (table 3.1), then since $\lambda \propto T/p$ at 2400 K and p is of order 10 Torr, $\lambda_W < d$ as expected. Also for the same value of relative evaporation λ_W in Ar is greater than λ_W in N_2 and hence the required p in Ar is lower than that in N_2, as is observed.

motion of charged particles in electric and magnetic fields

I shall treat this subject under three headings: the motion of charges *in vacuo*, in gases and in plasmas. No attempt has been made to give a comprehensive account, but I have selected some typical but mathematically simple problems with the aim of obtaining deeper physical insight rather than formal perfection.

4.1. Motion of charged particles *in vacuo*

Electrons and ions in electric fields

One of the simplest cases is the motion of an electron or ion in a uniform electric field in the absence of other charged or uncharged foreign particles or a solid wall. Let the mass of the charged particle be constant, its velocity small compared with that of light, and find the positions of the particle in an *x–y* plane as a function of the parameters of the motion.

Assume that a charge e is created at the point P (figure 4.1) when an electric field E in the x direction is present (remember that the motion of e (mass m) in a steady uniform electric field and that of an uncharged mass in a similar gravitational field are strictly analogous). Since the force on e is eE, it acquires in a constant electric field constant acceleration eE/m, its final velocity v_f after a time t, which for $v_0 = 0$ at $t = 0$, is $v_f = (e/m)Et$; the final linear momentum is $p = mv_f = (2emV_f)^{1/2}$ where V_f is the potential at $x = d$; the distance travelled is $d = \frac{1}{2}(e/m)Et^2$ and the time taken to cover d is $t = (2d/[e/m]E)^{1/2}$. The final kinetic energy, for $V = 0$ at $x = 0$, is $W = \frac{1}{2}mv_f^2 = eEd = eV_f$ and the action $A = Wt = p_f d = (2emV_f)^{1/2}d$.

If the charged particle were released from P at $t = 0$ with $v_x = v_0$ parallel to x and with E in the y direction, after a time t its y co-ordinate is

$$y = \tfrac{1}{2}(e/m)Et^2 \qquad (4.1)$$

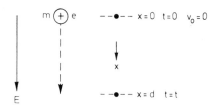

Figure 4.1. Non-relativistic motion of a charged mass m in a uniform field E.

and since $x = v_0 t$, by eliminating t, the trajectory is

$$y = \frac{(e/m)E}{2v_0^2} x^2 \tag{4.2}$$

which is a parabola. By comparing curves a and b in figure 4.2 we find that for a given x the distance y of the particle parallel to the field is larger the smaller the initial velocity v_0. The amount of energy taken by e from the field E is

$$W = \int_1^2 eE \, dy = e(V_1 - V_2) \tag{4.3}$$

Thus, the change in kinetic energy is proportional to the potential difference between the points 1 and 2 in figure 4.2. If v_0 is a constant, and the initial direction α is varied between $-90°$ and $+90°$, the parabolas c and d result, being the standard paths of a ballistic missile. From equation 4.2 it follows that the trajectory depends on e/m for given values of E and v_0, if v_0 is normal to E. The case of v_0 pointing in a general direction is left as an exercise to the reader.

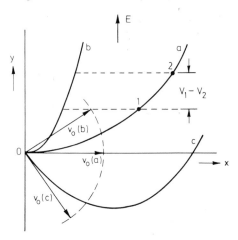

Figure 4.2. Motion in the $x - y$ plane of a charged mass (e, m) when $E = $ constant and its initial velocity v_0 points in different directions x.
$\alpha = 0$ for $V_0(a)$; points 1 and 2 indicate the positions in which $V_1 - V_2$ is the potential difference picked up by e, m.

The motion of a charge in a non-uniform electric field is fundamentally different from that in a uniform field where a charge of zero initial velocity moves along the straight field lines. Yet this is not so in a non-uniform field, when its field lines are curved. Hence a charge advancing along an element of path follows first the tangent of a line of force; the curvature of the line demands an acceleration normal to the tangent, i.e. a corresponding electric force component. However, the particle's inertia delays the radial motion, so the charge moves first an infinitesimal distance tangentially. Therefore, the trajectory of the particle runs across the field lines and the curvature of the trajectory is always smaller (its radius larger) than that of the field lines, as figure 4.3 illustrates.

This result can be amplified by an approximate calculation. Consider a charge (e, m) which moves in vacuum with a velocity v_0 in a general direction towards a stationary charge of opposite sign. Its trajectory, shown in figure 4.3, results from the attraction between the charges that turns the moving charge towards the stationary charge. To estimate the angle of deflection, let the distance of closest approach between the moving and fixed charge be r_0. Since the electric force between distant charges varies as $1/r^2$, substantial interaction will occur only over a path length of the order $2r_0$. The interaction time is therefore

$$\Delta t \simeq 2r_0/v_0 \tag{4.4}$$

The radial velocity component, i.e., its radial acceleration in Δt, is

$$v_r \simeq \tfrac{1}{2}(e^2/mr_0^2)(\Delta t) \tag{4.5}$$

and since $v_r \ll v_0$ the angle of deflection α is of the order v_r/v_0:

$$\alpha \simeq 2(e^2/mr_0 v_0^2) \simeq (e^2/r_0)/\tfrac{1}{2}mv_0^2 \tag{4.6}$$

The scattering angle thus depends on the ratio of the potential to the kinetic energy (see Chapter 3). We find, for example, from equation 4.6 that a 100 eV electron passing a positive ion at a (closest) distance of 100 atomic diameters (10 nm) is deflected by $\alpha \simeq 1\cdot4 \times 10^{-3}$ radian $\simeq 8 \times 10^{-2}$ degree.

If interaction occurs with many positive ions, an estimate of the average angular deflection can only be given if the number of such interactions or collisions at a distance is large. Let the ions be uniformly distributed throughout the space and the distance of closest approach between the moving electrons and

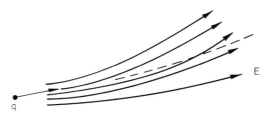

Figure 4.3. Trajectory of a charge q (dashed curve) of zero initial velocity in a non-uniform field E. Its motion across the diverging field lines is due to the inertia of q, m.

the fixed ions be always $r_0 = 10$ nm. Out of 1000 collisions, 500 will scatter the electron to the right and 500 to the left of its path. The actual deviation from its path is given by the statistical surplus of deflections to one side over that to the other side, which is equal to the square root of the total number of collisions, approximately 32. The mean square deflection after a thousand collisions is therefore $32 \times 0.08 = 2.6°$.

Consider now the case of an electron beam which passes from one field-free space into another when a potential difference across the boundary between the two regions is applied (figure 4.4). The potential difference ($V_2 - V_1$) between the two close metal electrodes (with holes at their centres) causes an abrupt deflection of the electron beam in the narrow space between the electrodes. This is because the horizontal velocity component v_{h_1} of the charge leaving region 1 is suddenly raised owing to the applied potential difference. From figure 4.4 it follows that, since the beam velocity $v \propto V^{1/2}$, $\tan \alpha = v_{v_1}/v_{h_1}$, $\tan \beta = v_{v_2}/v_{h_2}$ and $v_{v_1} = v_{v_2}$. The relation between angles and potentials is

$$\frac{\tan \alpha}{\tan \beta} = \frac{v_{h_2}}{v_{h_1}} = \left(\frac{v_2}{v_1} \right) \tag{4.7}$$

This equation is analogous to Snell's law, which relates the ratio of the refractive indices to that of the speed of light in two media.

Note that interchanging object and image has no effect on the path geometry. Hence, the path of the electron beam can be reversed without changing its refraction at the boundaries, provided the magnitude and polarity of the applied potentials remain unchanged. However, the conditions in the electric circuit change. When the beam proceeds from medium 1 at potential V_1 to 2 at

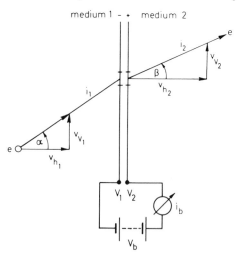

Figure 4.4. Trajectory of e from a field-free medium 1 at potential V_1 to medium 2 at V_2 separated by an infinitesimal gap bounded by plane conductors with a small hole. h = horizontal, v = vertical component of V; i_1, i_2 = beam current in 1 and 2.

potential V_2 (figure 4.4) the electrons are accelerated because the velocity component in the field direction is raised. The power P delivered by the battery of voltage V_b is, with

$$V_1 + V_b = V_2 \qquad (4.8)$$

and the electron current

$$i_1 \simeq i_2 = i_b \qquad (4.9)$$

$$P = V_b i_b = (V_2 - V_1)i_1 \qquad (4.10)$$

Thus, the battery supplies power to the beam although its electrons do not 'physically' strike the electrodes but pass through the holes. At first it seems to be odd that the electron current in the metal wires connecting the battery with the gap electrodes should flow across the gap without electrons being emitted from one and received by the other electrode. However, this problem will now be discussed in detail.

Suppose a charge q is moved from A to B (figure 4.5) through a vacuum or a gas-filled space bounded by two large parallel electrodes P and Q which are electrically connected. What causes the charge to move is at the moment irrelevant and so is the presence of a gas. When q is at A, more lines of force end on P than on Q, and the reverse holds when q is at B. At P (and Q) the induced charge changes with time at a rate

$$i_t = \frac{dq}{dt} = \frac{dq}{dx}v \qquad (4.11)$$

If the velocity v of q between A and B is constant, the current $i(t)$ will be a square pulse. The charge $\int i\,dt$ circulating in the circuit will be only a fraction of q but will be equal to q when A and B are at P and Q respectively. Thus, a continuous current can be maintained if, for example, q is injected at A at a large rate and removed at B, which is precisely the case when a beam of electrons passes the narrow gap in figure 4.4, thereby picking up energy in the field while falling through $V_2 - V_1$. A current between two electrodes is not therefore necessarily

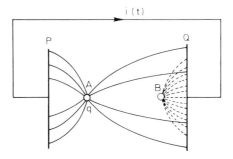

Figure 4.5. Field lines of a non-uniform electric field and path (dashed) of a charge q *in vacuo* which moves across the lines from A to B. A current pulse $i(t)$ ensues.

associated with charges which are emitted and collected by conducting electrodes.

A more general approach, illustrated in figure 4.6, is applicable to evacuated as well as to gas-filled systems. Assume that a single charge moves in a steady non-uniform electric field that is maintained by applying a voltage V_0 to electrodes of dissimilar shape. Equating the energy supplied to that used to move the charge e along dx we have

$$V_0 i_t \, dt = eE \, dx \cos \alpha \tag{4.12}$$

and thus

$$i_t = \left(\frac{E}{V_0}\right) e v_t \cos \alpha \tag{4.13}$$

Here i_t and v_t are the instantaneous current and velocity, and E/V_0 the 'topology factor' which describes the character of the electric field. From equation 4.13 we find that, for example, in a uniform field $\alpha = 0$, $E/V_0 = 1/d$, d being the electrode separation, whereas in a field between concentric cylinders r_1, r_2 the factor $E/V_0 = (r \ln r_2/r_1)^{-1}$ and $\alpha = 0$. Hence, $i_t = e v_t/d$ and $i_t = e v_t/(r \ln r_2/r_1)$ respectively. *In vacuo*, $v_t = f(V_x)$ as shown in Section 4.1, while in a gas v_t depends on E/p (Section 4.2) provided equilibrium is attained.

Neutral particles of known induced dipole moment flowing through a very strong non-uniform electric field produce a feeble current in the circuit. To calculate its value is a useful exercise for the reader.

The fact that electron beams are 'refracted' when passing a restricted region of high electric field (or potential difference across a narrow space) leads to the conclusion that 'electrostatic lenses' can be constructed. To bring a wide electron beam to a focus, a simple electron–optical system consisting of two uni-axial metal tubes can be used. Make a sketch of its field and equipotential distribution, the latter corresponding in geometrical optics to surfaces of constant refractive index. Trace the electron paths in the diverging beam section as well as its convergence towards the focus. The analysis, too unwieldy for our purpose, can

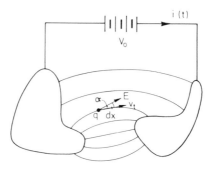

Figure 4.6. Current $i(t)$ circulating in a system when a charge q moves along dx in a field E set up by applying V_0 to two electrodes, v_+ being the instantaneous velocity and $E \, dx$, the angle α.

be found in special texts. So far we have treated only electrostatic fields set up by surface charges and described by Laplace's equation $\nabla^2 V = 0$ (in one dimension it reads $d^2 V/dx^2 = 0$). In the case of space charge fields, Poisson's equation holds, i.e., $\nabla^2 V = 4\pi\rho/\varepsilon_0$, ρ being the net space charge and ε_0 the electric permittivity. An example, in which an electron beam is deflected when crossing a space charge field *in vacuo*, will now be discussed.

Consider a finite region of uniformly distributed positive space charge ρ^+ of thickness d and width s in front of a plane negative electrode at $z = d$ (figure 4.7). A beam of fast electrons of velocity v_0 is shot across the space charge and its deflection Δ is measured on a fluorescent screen FS. Δ is a measure of the electric field E in the space charge region at the point of entry z_1, and by moving the point of entry, E can be found as a function of the distance $(d-z)$ from the cathode. Such space charge fields exist, for example, in the cathode dark space of a glow discharge. (The presence of gas has no effect on the magnitude of the deflection, but only on the spread of the beam, provided $v_0 = $ a constant throughout s.)

Let $z = 0$ be at the boundary between the dark space DS and the negative glow NG (figure 4.7). Since $\rho_e \ll \rho^+$ and $dE/dx = \rho^+/\varepsilon_0 = c_1$ we have

$$E = c_1 z \tag{4.14}$$

c_1 being a positive constant defined above and ρ^+ the net space charge.

Beam electrons of speed v_0 enter DS at z_1, cross the space charge of width s, and are accelerated to the right (by E) at a rate

$$\frac{d^2 z}{dt^2} = -c_2 z \tag{4.15}$$

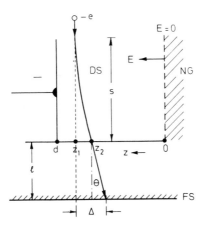

Figure 4.7. Measuring E in the dark space DS of a glow discharge at z, by observing the deflection Δ of an electron beam $-e$; Little and von Engel (1954).
NG = negative glow; FS = fluorescent screen.

where $c_2 = (e/m)c_1$. With $dz/dt = 0$ at $z = z_1$, we obtain

$$\frac{1}{c_2}\frac{dz}{dt} = -(z_1^2 - z^2)^{1/2} \tag{4.16}$$

By integrating once more, the time is obtained

$$\Delta t = s/v_0 = -\int_{z_1}^{z_2} \frac{dz}{[c_2(z_1^2 - z^2)]^{1/2}} \tag{4.17}$$

With $z = z_2$ at the point of exit the solution is

$$z_2/z_1 = \cos(c_2^{1/2}s/v_0) \tag{4.18}$$

z_2 is found from the deflection Δ (figure 4.7) given by $\Delta = (z_1 - z_2) + l\tan\theta$ where l is the screen distance and $\tan\theta = dz/dl = dz/dt\ 1/v_0$ which follows from equation 4.16 for $z = z_2$. Since $c_2 \propto c_1$, the absolute value of E can be found from the known electron beam energy through v_0.

So far I have only discussed trajectories of charges in constant electric fields. I shall now treat their motion in fields varying with time.

Consider a charge (e, m) which is subject to a uniform electric field E which varies sinusoidally with frequency f, namely $E = E_0 \sin\omega t$ where $\omega = 2\pi f$, which seems to be a simple problem. Assume that the field $e(t)$ starts to accelerate a charge e at $t = 0$ when the phase angle between E and the acceleration is ϕ, so that

$$\frac{d^2x}{dt^2} = \frac{eE_0}{m}\sin(\omega t + \phi) \tag{4.19}$$

If the charge created at $\omega t = 0$ has an initial speed v_0 in the field direction, then

$$\frac{dx}{dt} = v_0 + \frac{eE_0}{m\omega}[\cos\phi - \cos(\omega t + \phi)] \tag{4.20}$$

If the electron's position at birth is $x = 0$, the distance travelled is

$$x = \left[v_0 + \frac{eE_0}{m\omega}\cos\phi\right]t + \frac{eE_0}{m\omega^2}[\sin\phi - \sin(\omega t + \phi)] \tag{4.21}$$

This shows that the path is, in general, the sum of a sideways (drift) plus an oscillatory motion (figure 4.8). The former is due to the initial velocity v_0 (or its component parallel to E) and the phase angle ϕ with respect to the E.

The intuitive answer, that the charge's path is simply a straight line, along which it oscillates to and fro, is wrong. A pure oscillation would require the first term bracketed in equation 4.21 to vanish so that both v_0 and $\cos\phi$ were zero. In this case, the electron must start with $v_0 = 0$ when $E = E_0$. Yet only an infinitesimal number of electrons can be created during the interval dt at $\phi = \pi/2$. It is obvious that an electron subjected to $E(t)$ in an evacuated vessel will move towards the wall and finally impinge on it. From equations 4.20 and 4.21 it

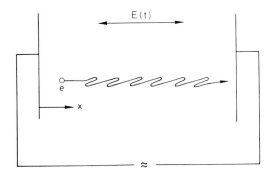

Figure 4.8. Path of charge e in an a.c. (or h.f.) electric field $E(t)$ released at $t=0$, $\omega t = \phi$.

follows that the speed and path amplitudes are smaller the larger the frequency, which is due to the inertia (mass) of the charge. The above equations hold for electron and ion oscillations. For example, a field $E_0 = 3000$ V m^{-1} at $f = 10^7$ Hz ($\omega = 2\pi 10^7$ s^{-1}) gives an electron a velocity amplitude $eE_0/m\omega$ of about 8×10^6 m s^{-1} (130 eV) and an oscillation amplitude of about 0·12 m. The corresponding values for an H$_2^+$ ion would be about 2×10^3 m s^{-1} and 30 μm respectively.

Electrons and ions in magnetic fields

If a charge q moves with velocity v into an evacuated space in which a constant magnetic field B is maintained, it represents a finite current–time element $\Delta(it)$. Since $t = d/v$ this is equivalent to a current–distance element $\Delta(id)$ of speed v that carries with it a magnetic field of its own. Therefore, when q moves in B, two magnetic fields interact (figure 4.9). When q moves in the direction v, the local magnetic field is strengthened on the right and reduced on the left. The redistributed field lines produce a lateral pressure difference (Maxwell)

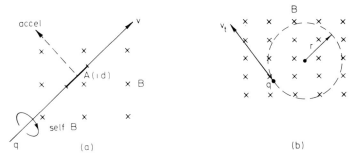

Figure 4.9. (a) Current element accelerated by a uniform magnetic field B *in vacuo*. (b) Charge moving with $v_t \perp B$ in a Larmor circle.

accelerating q to the left. It is seen that the deflecting force F is normal to both v and B, and is proportional to the vector product vB

$$F = q\mathbf{v} \wedge \mathbf{B} \tag{4.22}$$

No work is done by B on q because the path of q is normal to a circle around B. Thus $\int F \, dx = 0$. Both v and s, the centripetal acceleration due to F, remain constant. Since the v component parallel to the paper is also conserved, the path is a circle if v is in the plane (figure 4.9(b)) or a circular spiral when v is inclined at an angle to the plane of the paper.

Since the radius r of the circle is constant, F is balanced by the centrifugal force. If the charge q has mass m and $\mathbf{v} = \mathbf{v}_t$, v_t being the tangential velocity, the radial force is

$$mv_t^2/r = qv_t B \tag{4.23}$$

The radius of the circular path is therefore

$$r = \frac{mv_t}{qB} \tag{4.24}$$

For an electron one finds the 'Larmor radius' $r = 34 \, V^{1/2}/B$ (in mm) where V is its kinetic energy in eV and B is in Gauss ($1 \, \text{G} = 10^{-4} \, \text{T}$). The time taken for one circular path is $T = 2\pi r/v_t = 2\pi m/qB$, which is seen to be independent of v_t. The angular velocity ω of q, the 'Larmor frequency', in rad s^{-1} is

$$\omega = \frac{2\pi}{T} = \frac{qB}{m} \tag{4.25}$$

and is independent of v_t and r. It seems paradoxical that this should hold generally, say when $v_t \to 0$, but it makes sense when it is realized that now $T \to \infty$. By convention, r and ω are always taken as positive quantities. For a 100 eV electron ($v_t = 6 \times 10^6 \, \text{m s}^{-1}$) and $B = 10^{-2} \, \text{T}$, $r = 3 \cdot 4 \, \text{mm}$, $T \simeq 3 \, \text{ns}$ and $\omega = 2 \times 10^9 \, \text{s}^{-1}$. For a proton (H$^+$) of the same energy, we find $r = 2 \times 10^{-3} \, \text{mm}$, $\omega \simeq 3 \times 10^6 \, \text{s}^{-1}$. If H$^+$ has in addition $v_x \neq 0$ it appears to move simply along B, since the spiral radius is very small. Note that the circular motion of q (figure 4.9(b)) produces a self-field B that is opposed to the applied B (diamagnetic effect).

In a steady non-uniform magnetic field symmetrical about the x axis (figure 4.10) a positive charge q, injected at P normal to x and the paper, will find a field vector B which has z and r components, both perpendicular to the initial velocity v_0. B_x gives rise to an acceleration towards the x axis, B_r to an acceleration in the x direction, i.e., a force acting in the direction of decreasing B. Since equation 4.24 shows that an increase in B demands a decrease in r and equation 4.25 a rise in $\omega = v_t/r$, it follows that the path of q to the right will be along a helix of gradually increasing pitch. The same result is obtained by assuming q to be in a strong field so that it moves along a spiral of very small radius and thus follows

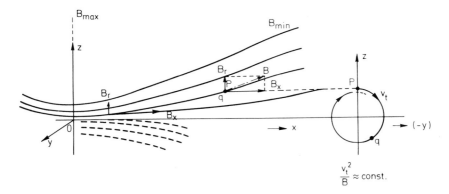

Figure 4.10. Motion of q in $B(x, r)$ along a spiral of varying radius; relating to "magnetic bottle, mirror and trap".

the field lines. If these lines are bent in the form of figure 4.10, q will follow them and is accelerated in the direction of x into the region of weaker field.

The resultant velocity v in B was said to be a constant of the motion. Yet in a general helical motion, we must distinguish between v_t and v_x, the tangential and axial (x) components of velocity, respectively. Conserving total energy $W_{total} = \frac{1}{2}m(v_t^2 + v_x^2)$ and diamagnetic moment M of the circling charge, namely $M = I \times A = $ filamentary current $I \times$ enclosed area A, where $I = q\omega/2\pi$, we find from equations 4.23 and 4.24 that

$$M = \tfrac{1}{2}q\omega r^2 = \tfrac{1}{2}mv_t^2/B \tag{4.26}$$

It can be shown that $M = $ constant holds for 'slow' changes of the field, i.e., if the change of B with x is accompanied by a large number of rotations of q about x, or if the charge q is uniformly distributed over the circumference of the circle. Therefore, a motion of the centre of the near-circular path of q to the left (figure 4.10) results in an increase in tangential velocity v_t. At the same time, the component force along x is obtained by differentiating equation 4.26 with respect to x (dt' being the time differential)

$$F_x = m\frac{dv_x}{dt'} = -M\left(\frac{dB}{dx}\right)_x \tag{4.27}$$

showing that as x becomes smaller, both the gradient of B and the repulsive negative force rise and v_x decreases. This demonstrates that W_{total}, the energy arising from v_x and v_t, does not change. Thus, the value of v_t reaches a maximum when $v_x = 0$. Whether or not this actually occurs depends on the geometry (see later).

Most of the arguments presented hold only when $v \ll c$ and when a small movement in the x direction is coupled with a long rotational path, i.e., with a large number of rotations around the field lines. Figure 4.10 relates to this case.

Let us turn to the case of $v_x \to 0$. If the angle between the velocity vector **v** and the x axis is α, we have

$$v_t = v \sin \alpha \tag{4.28}$$

Since M, given by equation 4.26, is constant it follows that

$$v_t^2 \propto B \tag{4.29}$$

and since the resultant velocity v is invariant, we find from equations 4.28 and 4.29

$$\sin^2 \alpha \propto B \tag{4.30}$$

A particle starting to gyrate at B_{min} at an angle $(B_{min}, v_t) = \alpha_{min}$ will be drifting in the $-x$ direction, i.e., the direction of rising B (figure 4.10). On arrival at B_{max}, where α_{max} is the angle between v_t and B_{max}, the relation between the two positions is given by

$$B_{max}/B_{min} = \sin^2 \alpha_{max}/\sin^2 \alpha_{min} \tag{4.31}$$

or

$$\sin \alpha_{max}/\sin \alpha_{min} = (B_{max}/B_{min})^{1/2} \tag{4.31a}$$

Note that B_{max} is not the position where B has the highest value in the system. Since $\sin \alpha_{max} < 1$, reflection of the gyrating charge or a change of sign of v_x can only occur when $\sin \alpha_{min} > |(B_{min}/B_{max})^{1/2}|$, whereby $\alpha_{max} = 90°$. For smaller values of α_{min} the charge is first retarded but later escapes into the region of negative x. Charges in fields of the configuration indicated in figure 4.10 can, therefore, under the conditions given by equation 4.31 be 'magnetically trapped'. The field region about B_{max} is called a magnetic mirror. Thus, two short energized coils form a pair of 'mirrors' between which the circling charges are oscillating (or drifting to and fro) along their common axis.

The relations derived above are, of course, valid for electrons and ions. It has been shown in equation 4.24 that a charge of a given tangential velocity v_t in a given field B travels on a circular path of (Larmor) radius r which is proportional to its mass m. Also, from equation 4.26, under the same assumptions, the magnetic moment M is proportional to m.

In table 4.1 the results obtained in uniform fields are compared with those derived for non-uniform fields.

Table 4.1. Comparison of uniform and non-uniform fields for v_t, r and ω.

B	v_t	r	ω	Remark
Uniform	$\propto Br$ which is independent of B	$\propto B^{-1}$	$\propto B$	$v_t = r\omega$
Non-uniform	$\propto B^{1/2}$	$\propto B^{-1/2}$	$\propto B$	

Electrons and ions in combined electric and magnetic fields

The path of a charged particle in an electric and magnetic field, both uniform and parallel to each other, is easily found. Assuming the charge starts with an initial velocity $v_\perp$, perpendicular to the fields. When E and B are in the same direction the path is a helix with a gradually increasing pitch but the pitch direction is reversed if E and B are in opposite directions.

If B points out of the paper and E is in the z direction, the path of a positive charge is the long-dashed curve in figure 4.11, namely a common cycloid. This motion arises only if the initial velocity of the charge is zero and the path starts at the origin. (The path is that of a point at the circumference of a wheel rolling over a plane.) The explanation of this motion is as follows. Originally the charge is accelerated by E in the z direction and at the same time in the x direction by B, i.e., perpendicular to $\mathbf{v} \wedge \mathbf{B}$. In spite of the component deflection along x, the particle speed increases, though its acceleration in E gradually falls since the potential it has fallen through is $\propto z$. When v has reached its maximum, after half a cycle, the magnetic force drives the particle in the direction $(-z)$ against the electric field. The energy which the particle has earlier drawn from the source of E is now gradually returned to it. This trend can obviously only persist until the speed v of the particle has reached zero value. Because of symmetry, this occurs when the particle reaches the x axis, whereupon it starts to move in the z direction, on its second cycle.

Observe: (a) that the direction of motion along x (but of course not along z) is independent of the sign of the charge; (b) that the time to complete a cycle is equal to the period given in equation 4.25; (c) that the rate of progress along x is greater the smaller the mass of the charge; and (d) that the path is a curtate or prolate cycloid (figure 4.11) if it starts with a finite initial velocity v_0 normal to B and E in the x–z plane.

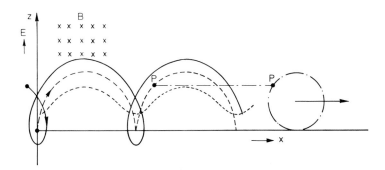

Figure 4.11. Motion of $+e$ ($v_0 = 0$) in $E \perp B$, described by point P on rolling circle. For $v_0 \neq 0$ the steady-state path is a cycloid, namely prolate when $0 < v_0 < 2u$, $u = E/H$ (see figure 4.12); common when $v_0 = 0$ (short-dashed); and curtate when $v_0 > 2u$ (solid). Otherwise transients in path have to be considered.

The superposition of a constant electric field and a magnetic field normal to it enables us to select from a beam of charged particles with distributed velocities those within a narrow range of velocities. Suppose the beam consists of charges which move with different velocities along x (figure 4.12) and enter a space with an E field pointing upwards (z) and a B field pointing into the paper (y). If only E would act on the charge $+q$ then the beam's curved path $y \propto x^2$ would bend upwards, whereas if B alone were present the beam would bend downwards and follow a circle. Therefore, in combined fields the electric force eE can be annulled by the electrodynamic force $(q) \mathbf{v} \wedge \mathbf{B}$, so that only charges of velocity

$$v_s = E/B \tag{4.32}$$

can move along a straight line *in vacuo*. Using a pair of apertures (figure 4.12) a beam of charges of uniform velocity can be separated from the original beam. This 'velocity-monochromator', therefore, does not separate charges or masses. To obtain, for example, electrons from a non-uniform beam of velocity $v_s = 6 \times 10^6$ m s^{-1} (100 eV), a field $B = 10^{-3}$ T (10 Gauss) and $E = 6$ kV m^{-1} must be provided.

The conclusion that as $B \to 0$ in equation 4.32 charges with $v_s \to \infty$ are selected is erroneous; equation 4.32 only holds as long as the parabolic path due to E and the circular path due to B 'annul each other'; this restricts the range of values of E and B which can be applied.

4.2. Motion of charged particles in gases

Consider the motion of a charged particle of finite initial kinetic energy in a gas at moderate density when electric and magnetic fields are absent. During its journey it will collide frequently, mostly elastically, with molecules so that its path is composed of a large number of short straight tracks pointing in different directions. In spite of this 'randomization' the charge is likely to proceed in the general direction of its initial velocity. However, its path in a uniform electric field consists of short curved tracks; after colliding, the velocity vector points in a

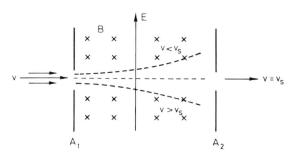

Figure 4.12. Velocity selector for a beam of charged masses.
Note that the selected velocity v_s in equation 4.32 is in S.I. units.

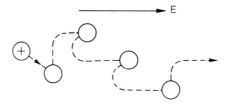

Figure 4.13. A charge drifting in a uniform field *E*.
Its path is composed of parabolic sections kinked by molecular collisions.

general direction, but the accelerating field tends to turn it into the field direction. This causes a 'drift' along the field plus superimposed erratic lateral movements (figure 4.13). The lateral scatter becomes larger the higher the gas density and the lower the particle's kinetic energy. If a magnetic field is applied, the radially accelerated charge tends to circle about the field lines, but in a gas, as a result of collisions, its path is composed of a large number of short kinked arcs (figure 4.14).

Motion of electrons and ions of given initial energy

Examine the motion of an electron which enters a field-free gas (V_i = ionization energy) with a kinetic energy $V_0 > V_i$ and thus $V_0 \gg kT_{gas}$ and observe it until $V \to kT_{gas}$. Assume it has gained V_0 by passing, without collision, through two closely spaced apertures, distance d apart ($d < \lambda$), that are kept at potentials 0 and V respectively (figure 4.15). Though little energy is abstracted in a single elastic collision ($\Delta V \ll V_i$, $\Delta V \ll V_0$, elastic scatter causes negligible gas heating) electrons of several hundred electron volts or less suffer a small number of energy-consuming inelastic collisions. As a result, new ions and electrons are produced pair-wise in quantities shown in table 4.2, as well as non-radiating metastable particles, radiating particles and dissociation products. The distance over which the primary electron is 'active' is called its range *R*, a poorly defined quantity because very slow electrons straggle and can induce vibrational and rotational excitation of molecules. The fact that, for example, 100 V electrons in rare gases (table 4.2) produce more ion pairs per electron in He than in Ne or Ar,

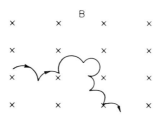

Figure 4.14. Path of an accelerated charge in a uniform field *B* consisting of circular arcs twisted by collisions.

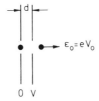

Figure 4.15. Producing an electron beam which enters a field-free gas through a holed aperture with energy eV.

may be due to the larger probability of excitation (compared to ionization) of the latter.

If the initial energy of the electrons V_0 exceeds the order of 1 keV, then about one half of it is spent on ionization, and the rest on the excitation and kinetic energy of secondary electrons in the gas. Slow electrons ($V_0 < 1$ keV) spend a smaller fraction of V_0 on ionization. Fast electrons produce about one ion pair for an average energy of 30–45 eV. Thus, a single 10 keV electron in nitrogen produces altogether about 300 ion pairs over its range and these are distributed along the path x as indicated in figure 4.16. This is due to the rise in excitation and ionization cross-sections with decreasing energy ε, except when ε is less than 100 eV (see figure 3.7).

Consider now slow positive ions of energy ε_0 less than 100 eV which move in their parent gas. Because ion and molecule have equal mass, the molecule can abstract on the average one half of the ion kinetic energy per collision. Therefore, the range of ions is in general small, and inelastic collisions of slow ions (except charge transfers) are often negligible. However, fast ions, such as α particles (He^{++}) of 5 MeV energy with a range of about 4 cm in atmospheric air, produce altogether approximately 10^5 ion pairs, whereas electrons of 5 MeV energy have a range of tens of metres and generate about the same total number.

Motion of ions through a gas in an electric field

Let a positive ion enter a gas in a uniform electric field. Assume that the ion has moved over a sufficiently large distance and has made many molecular collisions. Its motion is now in a steady state, i.e., the energy gained equals the loss, independent of position. As pointed out above, it is convenient to assign to the ion a mean random velocity v_i and, superimposed upon it, a drift velocity v_d.

Table 4.2. Total ionization I_{tot} (in ion pairs per electron) by electrons of initial energy ε_0.

Gas	He		Ne		Ar			Hg				H$_2$		N$_2$	
ε_0(eV)	50	100	75	100	30	50	100	30	50	100	200	100	200	75	100
I_{tot}	1·2	2·9	1·2	2·0	0·45	0·9	1·6	1·1	1·4	2·7	5·3	1·4	2·8	1·3	1·6

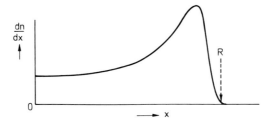

Figure 4.16. Local distribution $n(x)$ of the number of ion pairs cm^{-1} by 1 electron of 10 keV in N$_2$.

If the ions are in a sufficiently weak electric field E, that is if the mean energy ε^+ which they acquire is moderate, namely $v_d^+ \ll v_i^+$ or $E\lambda_i^+ \ll \varepsilon^+$, the ions display constant mobility. The drift velocity is then

$$v_d^+ = \mu^+ E \tag{4.33}$$

where μ is the ion mobility and E the electric field. At constant gas temperature T ($p = NkT$) this can also be expressed in similarity variables

$$v_d^+ = (\mu^+ p)(E/p) = (\mu^+ N)(E/N) \tag{4.34}$$

where N is the number density of gas molecules in m^{-3} and p the pressure in Torr. Note that the mean energy taken from the field is $E\lambda^+ \propto E/p \propto E/N$.

A mechanical formulation expresses the drift velocity in terms of a frictional constant c. Suppose a force f acts on an ion; its drift velocity in the gas is determined by a frictional force which in turn is proportional to the ion drift velocity v_d^+. By balancing electrical and frictional forces, using equation 4.33, and equating the second and fourth term in equation 4.35, we obtain

$$f = eE = cv_d^+ = c\mu^+ E \text{ and } c = e/\mu^+ \tag{4.35}$$

Thus, the frictional constant c is inversely proportional to the ion mobility.

Let us now express the ion mobility in terms of atomic constants. If m^+ is the mass of the positive ion, v_d^+ its drift velocity, ν^+ its collision frequency and τ^+ the constant time interval between two successive collisions, then by balancing the linear momentum we obtain

$$m^+ v_d^+ = f\tau^+ = eE \frac{\lambda^+}{\bar{v}^+} = m^+ \mu^+ E \tag{4.36}$$

where λ^+ is the mean free ion path, and $\bar{v}^+$ the ion random velocity, which is often approximately equal to that of the molecules. This equation shows that the mobility

$$\mu^+ \simeq \frac{e}{m^+} \frac{\lambda^+}{v^+} = \frac{e}{m^+} \tau^+ = \frac{e}{m^+} \frac{1}{\nu^+} \tag{4.37}$$

and since $\lambda^+ \propto \bar{p}^1$, $\mu^+ p$ is constant for constant T. Table 4.3 gives the mobility μ_1 of various ions in their parent gases at 1 Torr and 273 K. The mobility of

Table 4.3. Mobility μ_1 of atomic and molecular ions at 1 Torr and 273 K (observed).

Ion	Gas	μ_1 ($m^2\,V^{-1}\,s^{-1}$)	Ion	Gas	μ_1 ($m^2\,V^{-1}\,s^{-1}$)	Ion	Gas	μ_1 ($m^2\,V^{-1}\,s^{-1}$)
He^+	He	7.9×10^{-1}	H^+	H_2	1.1	$(air)^+$		1.4×10^{-1}
He_2^+	He	16×10^{-1}	H_2^+	H_2	1.0	$(air)^-$		1.9×10^{-1}
Ne^+	Ne	3.3×10^{-1}	H_3^+	H_2	>1.0	O_2^+	O_2	1×10^{-1}
N_2^+	Ne	5×10^{-1}	D^+	D_2	—	O_2^-	O_2	1.4×10^{-1}
Xe^+	Xe	4.4×10^{-2}	D_2^+	D_2	0.5	CO_2^+	CO_2	7.3×10^{-2}
Xe_2^+	Xe	6×10^{-2}	N_2^+	N_2	0.2			

negative ions is often about the same as that of positive ions, except in those gases which either form temporary negative ions (excited rare gases, N_2, H_2) or in which negative ions are continuously formed and destroyed (ions in air is such a case). Near ground level, molecular oxygen ions are present as a result of free electrons becoming attached and subsequently detached from O_2. The measured mobility of O_2^- is thus 40 per cent larger than that of O_2^+.

When the values calculated from equation 4.37 are compared with the data shown in table 4.3, it is found that the observed values are three to five times smaller than those predicted. This was originally thought to be due to 'clustering' (figure 4.17), which effectively increases the size and mass of the ion and reduces its mean free path. However, in the absence of impurities, this proved not to be the main cause. It was later found to be due to the interaction (change of linear momentum) of ions with electric dipoles, which they induce in the molecules of the gas (figure 4.18).

Consider molecules with zero permanent dipoles and let M_i be the induced electric dipole moment per molecule given by

$$M_i = \frac{k_d}{N} E_i = \frac{D-1}{4N} \frac{e}{r^2} \tag{4.38}$$

where D is the dielectric constant, k_d the dielectric susceptibility, E_i the field of the ion, r the distance from its centre and N the number density of molecules. For sufficiently large r, remembering the relation $H = 2M/r^3$ for the end-on magnetic field of a bar magnet, and using equation 4.38, we find the field E_D acting on the ion, owing to the induced dipoles in the gas, to be

$$E_D = \frac{2M_i}{r^2} = \frac{D-1}{2\pi N} \frac{e}{r^5} \tag{4.39}$$

This relation expresses E_D in terms of the dielectric constant of the gas (in excess of that of the vacuum) reduced to unit gas density. Note that E_D decreases

Figure 4.17. A cluster ion.

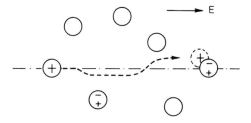

Figure 4.18. Momentum exchange between an ion and distant gas molecules caused by inducing electric dipoles in the latter.

with the inverse fifth power of the distance between the ion and the molecule in which the electric dipole has been induced. Since the average distance between two molecules is $r \approx N^{-1/3}$ and the mobility from equation 4.37 $\mu^+ = (e/m^+)\tau$, where $m^+ = m$, the collision time τ can be obtained by integration, thus

$$\tau = \int_0^R \frac{dr}{v_d} = \int_0^R \frac{dr}{\mu E_D} = \frac{2\pi N}{e\mu^+(D-1)} \int_0^R r^5 \, dr \approx \frac{1}{e\mu^+ N(D-1)} \qquad (4.40)$$

The gas density $\delta = mN$ and thus the mobility follows from

$$(\mu^+)^2 \propto [\delta(D-1)]^{-1} \text{ and } (D-1) \propto \delta \qquad (4.41)$$

$$\mu^+ = \frac{\text{constant}}{\delta} \left[\frac{1}{(D-1)_1} \right]^{1/2}$$

$(D-1)_1$ being the value of $(D-1)$ at $\delta = 1$.

Two points should be made: Firstly, whereas in equation 4.37 $\mu^+ p$ was found to be constant, this theory shows that $\mu^+ \delta$ is constant. Secondly, when equation 4.41 is compared with observations, a much better numerical agreement is obtained than before. The physical reason is, of course, that the inclusion of frequent small momentum exchanges with induced dipoles reduces considerably the collision time τ^+, and thus the mobility.

Positive ions drifting in an electric field through their own gas exhibit a 'mobility' determined by the 'charge exchange' (charge transfer) process. A helium atomic ion that collides with a He atom can be neutralized by extracting an electron from the atom. After separation the neutralized particle moves on with the velocity and in a direction it had before collision, whereas the newly born ion is being accelerated by the field (figure 4.19).

The charge exchange mobility can be treated as a 'stop–go' motion. Because the electron transfer requires the temporary formation of a compound molecule (re-arrangement of electrons), a close approach of ion and neutral is necessary. Acceleration of the ion mass M in the field E gives the local velocity $v(x)$ from

$$M \frac{d^2x}{dt^2} = Mv \frac{dv}{dx} = eE \qquad (4.42)$$

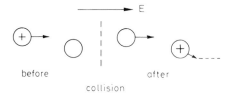

before   collision   after

Figure 4.19. Ion drift velocity in a parent gas at moderate E when charge transfer is the controlling process.

which for $v = 0$ at $x = 0$ yields

$$v = Ax^{1/2}$$
$$A = [(2e/M)E]^{1/2}$$

(4.43)

and a drift velocity averaged between 0 and λ_{+0}, the constant mean free path for charge exchange, is

$$v_d^+ = \frac{1}{\lambda + 0} \int_0^{\lambda+0} v \, dx = \left(\frac{8}{9} \frac{e}{M} E\lambda_{+0} \right)^{1/2} \propto (E/p)^{1/2}$$

(4.44)

a result similar to equation 4.46. By allowing λ_{+0} to be distributed, one finds the numerical term reduced to $(\pi/8)^{1/2}$. In a highly ionized gas, ion–ion interactions must be considered, which somewhat reduces λ_{+0}, but in a completely ionized gas with no neutrals no charge exchange occurs. The mean free path $\lambda_{+0} = (q_{+0}N)^{-1} = f(\varepsilon)$ is seen to increase as ε rises (figure 4.20), where q_{+0} is the charge transfer cross-section whose dependence on ε is small here. When the ion in question changes from He^+ to He_2^+, it is found that the mobility of the molecular ion, in spite of having twice the mass of the atom, is about twice as large. This shows that we are dealing here with a resonance effect, since virtually

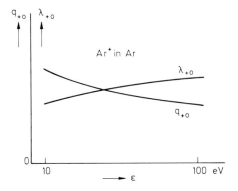

Figure 4.20. Dependence of the charge transfer mean free path $\lambda_{+0} = f(\varepsilon^+)$.

no momentum and energy is exchanged in the He^+ collision. Whenever dissimilar particles interact, such as He_2^+ in He, charge exchange is less probable and q_{+0} small.

Ions moving in a field of medium strength but not in their own gas show experimentally that their drift velocity is proportional to $(E/p)^{1/2}$. The physical reason is that the assumption of a constant collision time τ is not valid any more, since in higher fields the 'ion temperature' T_i (or $\bar{v}_i$) is greater than T_{gas} and thus τ rises with increasing E/p. Moreover, the drift velocity is no longer small compared with the random velocity of the ions. This result can be derived by allowing for a constant fraction $K = M/2M =$ one half of the mean energy lost in a single head-on collision. By balancing the energy per second gained from the field and the loss in kinetic energy that a single ion suffers, we find

$$eEv_d^+ = K^1 \tfrac{1}{2} M^+ (\bar{v}^+)^2 \bar{v}^+ / \lambda^+ \qquad (4.45)$$

Substituting $\bar{v}^+$ from equation 4.37 in equation 4.45, we obtain for $T_{gas} =$ constant

$$v_d^+ = \left[\frac{K}{2} \left(\frac{e}{M^+} \right)^2 \right]^{1/4} (E\lambda^+)^{1/2} \propto (E/p)^{1/2} \qquad (4.46)$$

We conclude that at low E/p, $v_d^+ \propto E/p$, whereas for a larger values of E/p, $v_d^+ \propto (E/p)^{1/2}$. Figure 4.21 shows the dependence of the positive ion drift velocity v_d^+ in noble gases as a function of E/p, the field per unit pressure (gas density), on linear scales, and figure 4.22 shows the drift velocity of atomic and molecular ions in their own gas; figure 4.21 is plotted on a double logarithmic scale to cover a large range of fields and velocities.

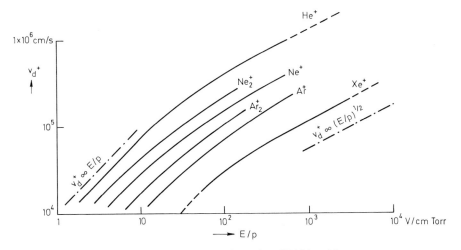

Figure 4.21. Atomic and molecular ion drift velocities $v_d^+ = f(E/p)$ in noble gases. Slopes of dash-dotted lines indicate proportionalities.

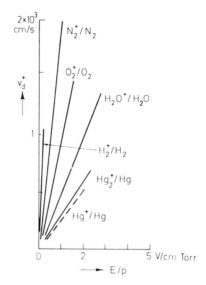

Figure 4.22. Ion drift velocity in molecular gases $v_d^+ = f(E/p)$.
At large p (order 1 atm) ions often associate with excited molecules.

The drift velocity of electrons and their mobility

When electrons move through a gas in a uniform electric field E, their movement at moderate E (figure 4.23) is akin to that of a swarm of bees. Electrons in a swarm have a velocity distribution, move in random directions and have a small component of velocity, the drift velocity v_d parallel to the field direction. Remember that the essential difference between electron and ion swarms is the mass ratio and the possibility of charge exchange. Even in relatively moderate electric fields, the average energy which an electron picks up along a mean free path, is large compared with the mean energy of the gas

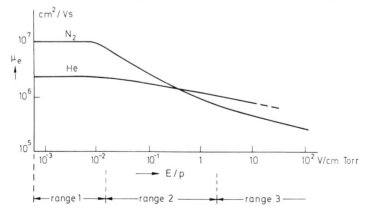

Figure 4.23. Electron mobility $\mu_e = f(E/p)$ in N_2 and He at low pressure and at about 300 K.

molecules ($eE\lambda_e > kT_{gas}$ and $T_e \gg T_{gas}$) and therefore they acquire a random velocity v_e which is considerably above that of v_d. This means that the mean collision time τ depends on the energy gain $E\lambda_e$ over a mean free path. Hence, a rise in $E\lambda_e$ and T_e is the main cause of the rapid fall in electron mobility with increasing field per unit gas pressure, E/p. Figure 4.23 gives the variation of the electron mobility μ_e in helium and nitrogen. Thus, except for range 1, where weak fields E/p mean that the energy acquired between two collisions is less than the gas kinetic energy (order 0.03 eV), a strong decrease in mobility occurs. This makes the mobility concept sometimes less meaningful.

It is therefore more convenient to plot the drift velocity v_d of electrons in He and N_2 as a function of E/p, the field per unit pressure, as figure 4.24 shows. In all molecular gases the curves, though never straight lines, rise linearly apart from a few small undulations. The reason is, of course, that the coefficient c_e in equation 4.48, and thus κ, is not a constant. The physical interpretation is that excitation of rotational and vibrational and, at higher fields, of electronic levels occurs when electrons interact with molecules. A more detailed study should disclose that undulations in the drift velocity curve are closely linked with the excitation of non-radiating states and optical spectra which are causing energy losses. It is seen that in range 2, figure 4.24, we can write $v_d = \text{constant} \times (E/p)^{1/2}$. This result is based on a simple physical concept.

As for ions, the electron drift velocity in larger electric fields is found by balancing the energy per second gained by an electron from the field against that spent by electron–molecule collisions at a rate of $\bar{v}_e/\lambda_e$ times per second, when $\kappa\varepsilon = \frac{1}{2}\kappa m \bar{v}_e^2$ is the mean energy loss per collision, κ being the constant fraction of energy lost. Hence, in equilibrium

$$eEv_d = \kappa\tfrac{1}{2}m\bar{v}_e^2\bar{v}_e/\lambda_e \tag{4.47}$$

and the drift velocity, by the procedure applied in equation 4.45, is

$$v_d = c_e(E/p)^{1/2} \tag{4.48}$$

where c_e is the electron mobility coefficient. This agreement between theory

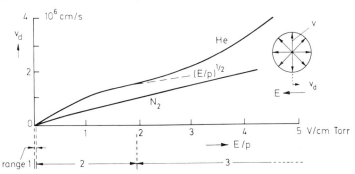

Figure 4.24. Electron drift velocity $v_d = f(E/p)$ in N_2 and He.
The curve for H_2 (not shown) lies about 15 per cent below N_2.

(equation 4.48) and observations only holds as long as elastic collisions predominate (range 2, figure 4.24). However, when the field is further increased, excitations and other inelastic collisions take place so that κ and v_d increase with the field at a faster rate than before (range 3, figure 4.24).

The physical reason is easily appreciated. Those electrons which lose large amounts of energy fall back into the lower energy range of the distribution, i.e., their group is 'cooled down'. Thus, suddenly deprived of this large random velocity, they are accelerated in the field direction and, though often scattered, apparently show a larger drift velocity than high-energy electrons in the distribution which suffer strong scatter, often in a direction opposite to that of the field. It follows that the 'colder' electrons drift faster, overtake the 'hot' group and arrive earlier at the end of the drift space. The observed (bulk) drift velocity v_d thus appears to rise fast with E^n ($n > 1$), the higher the loss factor κ and the frequency of inelastic encounters.

The average energy loss factor $\bar{\kappa}$ rises as E/p or kT_e is increased. In atomic gases and metal vapours $\bar{\kappa} \sim 2m/M$ at low T_e when elastic collisions predominate, whereas in molecular gases $\bar{\kappa}$ begins to rise when $T_e > T_{gas}$, because electrons excite molecules to rotational levels. Figure 4.25 shows approximately the variation of $\bar{\kappa}$ with kT_e.

Diffusion of charges

Charged particles diffuse through a gas when their number density N varies with position. The direction of diffusion is from the high-density to the low-density region and the flow density of charge is

$$F = -D\nabla N = -N(x)v_D \qquad (4.49)$$

where F is in particles m^{-2}, s^{-1}, v_D is the diffusion flow velocity in $m\ s^{-1}$ and $v_D = -(D/N)\nabla N$; the diffusion coefficient here is either D, D^+, D^- or D_e for neutral particles, positive ions, negative ions and electrons, respectively. Values for neutrals and ions are given in table 4.4. Owing to the differences in the nature of the collision processes the absolute values of the mean free path λ (or the related cross-section q) vary from table to table depending on the date and the method of measurement. The (self-) diffusion coefficient D of neutrals with the collision frequency per particle $\nu = \bar{v}/\lambda = 1/\tau$ (in s^{-1}), where τ is the collision time, is

$$D = \frac{\bar{v}\lambda}{3} = \frac{\bar{v}}{3qN} = \frac{\bar{v}^2}{3\nu} \qquad (4.50)$$

the mean velocity $\bar{v} = [(8/\pi)kT/M]^{1/2}$, λ is derived from viscosity data for neutral particles and from electron–molecule total cross-sections q_e for electrons. I shall

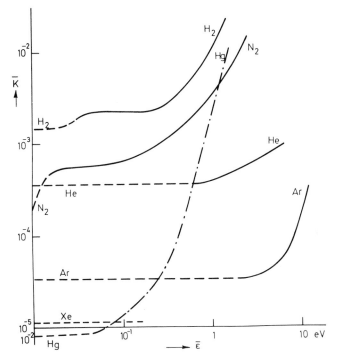

Figure 4.25. Electron energy loss factor $\kappa = f(\bar{\varepsilon})$ for various gases.
For $T_e > T_{gas}$, $\kappa_{elast} \sim 2\cdot6\,m/M$; for $T_e = T_{gas}$, $\kappa_{elast} = 2m/M$ whose value (in 10^{-6}) is 280, 56, 28, 13, 8·6 and 5·5 for He, Ne, Ar, Kr, Xe and Hg respectively. $\bar{\varepsilon} = 10^{-2}\,eV$ at the origin.

assume here that atomic and molecular ions (and excited particles) move through their parent gas of number density N. Negative ions may undergo detaching collisions so that for a part of the time the electrons are free until they are reattached to neutrals. In the case of positive ions, charge transfer may alter the coefficient of diffusion.

An important parameter (attributed to Einstein) is D/μ, the ratio of diffusion coefficient to the mobility. It is applicable to ions and electrons and holds as long as the coefficients are constant.

Table 4.4. Self-diffusion coefficient D_1 and ion diffusion coefficient D_1^+ at 300 K and 1 Torr pressure (in $cm^2\,s^{-1}$).

Gas	H_2	O_2	N_2	He	Ne	Ar	Kr	Xe	CO_2	Hg
D_1	1120	160	160	1200	400	140	73	42	87	~17
D_1^+	98	21	23	380	120	47	17	12		12

D_1 is taken from diffusion measurements, in general unreliable.
$D_1 \propto T^2$ approx.

With

$$\mu_e^+ = \frac{e}{m_e^+} \frac{\lambda_e^+}{\bar{v}_e^+} = \frac{e}{m_e^+} \frac{1}{v_e^+} \tag{4.51}$$

we find for e or + (electrons or positive ions singly charged)

$$\frac{D_e^+}{\mu_e^+} = \frac{m_e^+ (v_e^+)^2}{2} \frac{2}{3e} = \frac{\frac{3}{2} k T_e^+}{e} \frac{2}{3} = \frac{k T_e^+}{e} \tag{4.52}$$

Since D is in $m^2 \, s^{-1}$ and μ in $m^2 \, V^{-1} \, s^{-1}$, kT/e is in V. D/μ is often termed the 'characteristic temperature' or energy of the species and its value becomes meaningless either when E/N becomes so large that the energy distribution strongly deviates from the assumed one or when q_e varies with the mean electron energy $\bar{\varepsilon} = f(E/N)$.

So far I have tacitly assumed that singly charged species of one polarity disperse in the gas ('free diffusion'). Yet more often diffusion of charges of both polarities occurs, giving rise to an electric field. Coupling between unlike charges then takes place, particularly when the number density (concentration) reaches values that are encountered in discharges. In this case the movement of the electrons is not any more independent of that of the ions, but both tend to move together in the same direction with a common 'velocity'. This is called 'ambipolar diffusion' and is described by relations which hold provided the local charge concentrations N_e and N^+ as well as their gradients are equal.

The ambipolar diffusion coefficient D_a depends on the mean electron energy kT_e and can be derived as follows. Suppose there exists a common ambipolar velocity v_a so that

$$\mathbf{v}_a = \mathbf{v}^+ = \mathbf{v}_e \tag{4.53}$$

and the velocity component in the field direction x is given by the appropriate mobility and concentration gradient, then

$$v_a = \frac{D^+}{N^+} \frac{dN^+}{dx} + \mu^+ E = -\frac{D_e}{N_e} \frac{dN_e}{dx} - \mu_e E \tag{4.54}$$

Eliminating E and putting $N^+ = N_e = N$ and thus

$$\frac{dN^+}{dx} = \frac{dN_e}{dx} = \frac{dN}{dx}$$

we obtain

$$v_a = -\left[\frac{D^+ \mu_e + D_e \mu^+}{\mu_e + \mu^+} \right] \frac{1}{N} \frac{dN}{dx} = -\frac{D_a}{N} \frac{dN}{dx} \tag{4.55}$$

where D_a is the bracketed expression. Equation 4.55 shows that D_a is a diffusion coefficient weighted in the ratio of the mobilities. When $T_e \gg T^+$, since $\mu_e \gg \mu^+$, then $D_a \simeq D^+ + D_e \mu^+/\mu_e \simeq D_e \mu^+/\mu_e$. This result, combined with equation 4.52, gives

$$D_a = \frac{kT_e}{e} \mu^+ \tag{4.56}$$

showing that D_a is determined by the electron mean energy and the ion mobility, i.e., the nature of the gas. In an isothermal plasma $T_e = T^+ = T$, and therefore from the above, $D_a = 2(kT/e)\mu^+$. The field E can be easily found from equation 4.54.

Of considerable interest is the extent of the transition region from free electron diffusion (D_e) to the ambipolar one (D_a). Figure 4.26 shows the effective coefficient $D_{\text{eff}} = f(N_e)$ when kT_e, μ^+ and the tube radius R are given. The abscissa can also be given in units of $N_e R^2/kT_e$, a quantity proportional to $(R/\lambda_D)^2$ which follows from equation 4.75, λ_D being the Debye length. At large values of N_e, when the plasma approaches 'full' ionization but exhibits non-uniform number density of charge, 'plasma self-diffusion' develops with a diffusion coefficient D_{pl}. Its temperature dependence follows, for example, from the relation between the thermal conductivity κ and the diffusion coefficient D of gases, i.e., from the transport equations

$$\kappa = D\rho c_v \tag{4.57}$$

where ρ is the gas density $(\propto p/T)$ and c_v is the (electron–ion gas) specific heat (almost constant). One finds with $D = D_{\text{pl}}$ that

$$D_{\text{pl}} = \kappa(\rho c_v) \propto T^{7/2}/p \tag{4.58}$$

Like other diffusion coefficients $D_{\text{pl}}p$ is constant for constant T. In thermodynamic equilibrium $N_e = f(T)$, which explains the second falling part of the curve in figure 4.26. Plasma self-diffusion is seen to occur at a lower rate than ambipolar diffusion.

It is useful to consider under what conditions the concept of diffusion will fail if self-diffusion of neutral particles occurs. Obviously, diffusion ceases when scattering in the gas becomes scarce. Take a closed vessel of volume V, and radius r, and define 'scarce' by the ratio of wall:volume collisions, namely $(N\bar{v}/4)[4\pi r^2/(N\bar{v}/\lambda)] \frac{4}{8}\pi r^3 \simeq \lambda/r \simeq \lambda_1/pr$. With $\lambda_1 = 5 \times 10^{-5}$ m for a typical gas in a vessel of $r = 5$ cm, the collision ratio is unity when p is about 1 mTorr; at lower p instead of diffusive motion wall-controlled motion sets in.

Ambipolar diffusion in a magnetic field

When a cylindrical plasma column of radius R in a gas at low p is acted upon by an axial magnetic field B (figure 4.14) ambipolar diffusion of charges to the wall of the tube is impeded. This is because the electrons—which diffuse in $B = 0$

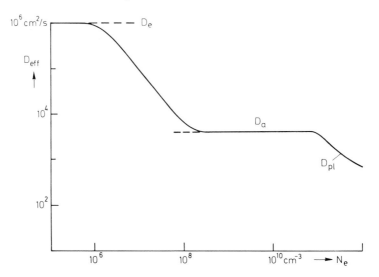

Figure 4.26. 'Effective' diffusion coefficient $D_{\text{eff}} = f(N_e R^2/kT_e)$ dependent on the electron concentration N_e of a cylindrical plasma in N_2 ($\mu_1^+ = 2 \times 10^3$ cm^2 (Vs)$^{-1}$) illustrating free electron (D_e), ambipolar (D_a) and plasma (D_{pl}) diffusion. $kT_e = 2$ eV, $R = 2 \cdot 4$ cm. Abscissa: $\alpha \Lambda_0^2 N_e/kT_e \propto (\Lambda_0/R)^2$.

radially outwards—now experience a tangential acceleration $(e/m)vB$, thus moving clockwise around B along curved paths, while the anti-clockwise acceleration of the slow heavy ions, though coupled to the electrons, does not affect the electrons' path. The resulting reduction in number losses is reflected in diminished values of T_e, of the radial electric field E_r, of the axial electric field E and of the (negative) wall potential. However, from equation 4.55, with $T_e \gg T^+$ and $\mu_e \gg \mu^+$, we have

$$D_a \simeq D^+ (T_e/T^+) = D_e(\mu^+/\mu_e) \qquad (4.59)$$

and thus a decrease of D_e and T_e by B reduces D_a. (Other effects of B on the plasma column are changes in radial distribution of charge, in wall potential, etc., points that will not be discussed here.)

If we consider the diffusion of electrons in a cylindrical ionized gas subjected to an axial magnetic field B, we can relate their radial drift velocity v_r to the driving force f by invoking the linear momentum $mv_r = f\tau$, τ being the collision interval in seconds. The Lorentz force $f = eBv_\perp$, where $v_\perp$ is v_t for the radial component (f_r) and v_r for the tangential component (f_t) of the force. Hence the radial velocity is the sum of diffusion and Lorentz velocity

$$v_r = -\frac{D_0}{N}\frac{dN}{dr} - eBv_t\frac{\tau}{m} = -\frac{D_0}{N}\frac{dN}{dr} - \left[\frac{eB\tau}{m}\right]^2 v_r \qquad (4.60)$$

where D_0 is D_e for $B = 0$.

With $\omega = eB/m$ from equation 4.25 we find, for constant T_e, that the diffusion coefficient D_B in a field B, with $\tau = 1/\nu_e = \lambda_e/v_e \propto 1/p$, is

$$\frac{D_B}{D_0} = \frac{1}{1 + (\omega\tau)^2} \lesssim (B/p)^{-2} \qquad (4.61)$$

The decrease of D_e with B is more involved when B affects not only ω but also τ, as shown earlier. A more complex situation arises in highly ionized plasmas and at large values of B where plasma fluctuations reduce the dependence on B, thus allowing plasmas to diffuse across B fields with greater ease than expressed in equation 4.61.

Self-repulsion of charges

Consider a spherical volume containing charges of like sign and uniform density ρ *in vacuo*. If the radius of the sphere is r_0 at $t = 0$, then each elementary shell of radius r and thickness dr is acted upon by radial force $df_r = 4\pi r^2 \, dr\rho_r E_r$ which will radially accelerate the charged masses. The ensuing motion is due to self-repulsion of charges. f_r is the total force which would arise if the total volume charge in the sphere was attracted by an equal uniform surface charge of opposite sign at infinity. Let e and m be the charge and mass of each individual particle. Gauss's law states that the electric flux ϕ_r entering a spherical shell minus that leaving it must equal 4π times the total charge enclosed. We therefore have

$$\phi_r - \phi_{r-dr} = \left\{ E_r r^2 - \left(E_r + \frac{dE}{dr} \, dr \right)(r + dr) \right\} = 4\pi r^2 \, dr\rho/\varepsilon_0 \qquad (4.62)$$

and thus for $0 < r < r_0$

$$\frac{1}{r^2} \frac{d(E_r r^2)}{dr} = \rho/\varepsilon_0 \qquad (4.63)$$

and for $r > r_0$

$$E_r = Q/(r^2 4\pi\varepsilon_0) \qquad (4.64)$$

Assume zero loss of charge and that Q is the total charge enclosed by the sphere of radius r_0 at $t = 0$. Consider now only ions at the sphere's surface. Since, at $t > t_0$, each ion at r is radially accelerated, we can write

$$\frac{d^2r}{dt^2} = \frac{e}{m} E_r = \frac{eQ}{mr^2 4\pi\varepsilon_0} \qquad (4.65)$$

Thus

$$r^2 \frac{d^2r}{dt^2} = \text{constant}$$

and with $dr/dt = 0$ for $r = r_0$ at $t = 0$, we find with $x = r/r_0$

$$\left(\frac{2a}{r_0^3}\right)^{1/2} t = \int_1^x \left(\frac{x}{x-1}\right)^{1/2} dx \tag{4.66}$$

where $a = eQ/m4\pi\varepsilon_0$; $(r_0^3/2a)^{1/2}$ is a characteristic time and x, the reduced radius. Finally we obtain

$$\left(\frac{2a}{r_0^3}\right)^{1/2} t = [x(x-1)]^{1/2} + \tanh^{-1}\left(\frac{x}{x-1}\right)^{1/2} \tag{4.67}$$

From equation 4.64 it follows that since $Er^2 = \text{constant}$, $\rho_r < 1/r^2$, indicating the decrease in charge concentration with distance. The critical reader may argue that for constant Q the potential difference

$$V - V_0 = Q\left(\frac{1}{r_0} - \frac{1}{r}\right)$$

corresponds to the kinetic energy $\frac{1}{2}mv^2$ which the charge e acquires *in vacuo* at r. This gives $v = f(r)$ and integration, to find $r = f(t)$, leads again to equation 4.67.

If the charges Q in the spherical volume, instead of *in vacuo*, are released in a gas at pressure p, the self-electric field of Q will drive those on the surface with a velocity approximately given by

$$v = \mu E_r = \mu Q/r^2 \tag{4.68}$$

where μ is the mobility, and

$$r^3 - r_0^3 = 3\mu Qt \propto t/p \tag{4.69}$$

using the same boundary conditions as in equation 4.66. We conclude that in a gas at $r > r_0$, $r \propto (t/p)^{1/3}$, whereas *in vacuo* (equation 4.66) $r \propto (t/m)^{1/2}$ holds.

A numerical example illustrates these results. Assume that a sphere of $r_0 = 5$ mm contains a charge $Q = (5 \times 10^8)(1 \cdot 6 \times 10^{-19}) = 8 \times 10^{-11}$ C of positive ions each of mass 5×10^{-23} g and average mobility $\mu^+ = 0 \cdot 5$ m^2 V^{-1} s^{-1} at 1 Torr. Then the time taken for the outermost ions to reach $r/r_0 = 10$ is, from equation 4.69, $t \simeq 3 \times 10^{-4}$ s. If the same charges repel mutually *in vacuo*, the time, from equation 4.67 is only $t \simeq 2 \times 10^{-6}$ s.

4.3. Motion of charged particles in plasmas

In Sections 4.1 and 4.2 the motion of electric charges *in vacuo* was shown to be controlled by external fields, whereas in gases it is influenced by collisions or short-range forces between charges and neutrals. The motion of charges in plasmas, however, depends on factors of an entirely different kind—the long-range forces. I shall try to clarify these 'collective' phenomena by an example.

When the movements of charged particles are considered it is useful to introduce their average random path length. In order to describe this path we shall study the motion of a single 'test particle' in its environment—the sea of charges which constitute a plasma. To simplify the problems I have chosen a quasi-neutral, fully ionized plasma, i.e., one containing zero average net space charge. The results can be easily comprehended. A particular form is the fully ionized hydrogen plasma which comprises equal numbers of protons (H^+) and electrons, implying that neutral atoms or molecules are absent.

The Debye length

Suppose electrons and positive ions in equal and large numbers occupy a volume of plasma of λ^3 where λ is a mean free path to be defined. If the electron and ion temperatures are equal and low, practically random motion occurs and thus unlike charges take up positions at equal distances from one another as ions in a solid lattice. However, at finite temperatures equipartition of energy causes the electrons (because of their small mass) to acquire considerable speeds. They fill regions near and around the ions and form a covering or negative charge which is subject to fluctuations as individual electrons enter and leave these regions. Thus, the electrons 'screen' the positive charge, so that an external (applied) electric field exerts a smaller force on the ions than when the electron screen is absent. This shows that the once uniformly distributed electron space charge is now redistributed so that its time average becomes more concentrated around the positive charges. This redistribution changes locally the field and the potential.

In the absence of other charges at finite distances r

$$V(r) = \frac{e}{4\pi\varepsilon_0 r} \tag{4.70}$$

which is the Coulomb potential of a single unscreened charge e. In the presence of a space charge ρ_{net} we have to solve Poisson's equation to obtain $V(r)$ of a screened charge. For spherical geometry, this reads

$$\frac{1}{r^2} \frac{d}{dr}\left(r^2 \frac{dV}{dr}\right) = -\rho_{net}/\varepsilon_0 \tag{4.71}$$

Let the ion temperature be T and the electron temperature be $T_e \gg T$. For small r—inside the screen—$V(r)$ approaches equation 4.70, whereas for large r, $V(r) \to 0$. Though ρ_{net} in the plasma, averaged over its whole volume, is zero or $(N^+)_0 = (N_e)_0$, we have around a positive ion at $r=0$ for a constant density of positives and a Boltzmann distribution of electrons

$$\rho_{net} = \rho^+ - \rho_e = (eN^+)_0 - (eN)_0 \exp\left(eV/kT_e\right) \tag{4.72}$$

where the exponential indicates that $N_e(V)$ results from a competition between the electron kinetic energy, given by T_e, and the potential energy, given by V. Expanding equation 4.72, i.e., for small values of eV/kT_e, we have

$$\rho_{net} = (eN_e)_0 [1 - \exp(-eV/kT_e)] \approx \frac{eV}{kT_e} (eN_e)_0 \qquad (4.73)$$

Inserting this in equation (4.71) gives

$$\frac{1}{r^2} \frac{d}{dr} \left(r^2 \frac{dV}{dr} \right) = \left[\frac{e^2 (N_e)_0}{kT_e \varepsilon_0} \right] V \qquad (4.74)$$

where $[\ \ ]$ is the inverse square of a characteristic distance Λ. Λ is the Debye or screening radius (length) which for $(N_e)_0 = N_e$ is

$$\Lambda = \left(\frac{kT_e \varepsilon_0}{e^2 N_e} \right)^{1/2} \simeq 70 \left(\frac{T_e(K)}{N_e(m^{-3})} \right)^{1/2} \qquad (4.75)$$

Thus, for $kT_e = 1 \text{ eV} = 11\,600 \text{ K}$ and $N_e = 1 \times 10^{16}$ charges m^{-3} at the screening distance Λ the potential is $1/e$ of the Coulomb potential given by equation 4.70, and $\Lambda \simeq 7 \times 10^{-5}$ m. If this is compared, for example, with the mean distance $\bar{d}$ between two charges (here $\bar{d} = N^{-1/3} \simeq 5 \times 10^{-6}$ m), it is seen that $\Lambda > \bar{d}$, i.e., the screened volume $N_e \Lambda^3$, contains a large number of charges (3500), as anticipated.

A solution of equation 4.74 with the above boundary conditions is one of the type e^{-r}/r, as can be seen by inserting equation 4.70 into equation 4.74. Thus, the 'screened potential' is

$$V(r) = A \frac{e}{r} \exp(-r/\Lambda) \qquad (4.76)$$

Comparison with equation 4.70 shows that the exponential term describes the screening effect of the ion by the electrons which are responsible for the rapid fall of V with rising r. Hence, most of the field lines of a central positive charge end in the electronic charges surrounding it and only relatively few lines extend beyond Λ.

The microfield

Let us discuss the scattering of a moving charge traversing a plasma while it experiences a large number of distant encounters. The instantaneous magnitude of the local electric field $E(t)$, randomly changing in direction, at a specified position r may be considered to be composed of an averaged field and, superimposed upon it, an internal fluctuating electric field $E(t)$ which is caused by the random motion of the large number of charges in the plasma.

I must underline that the random fluctuations of E_i occur very quickly and the corresponding high frequencies cover a fairly wide band; also, the field direction at a point varies quickly and irregularly. These electric 'microfields', short pulses with huge peaks, reveal their presence spectroscopically: line splitting by the 'internal Stark effect' is observed. It is obvious that the mean square microfield value is relatively low and the mean linear value must be zero by definition.

Approximate values of the microfield can be found by considering a sphere of radius R in which there are N_e and N^+ charges m^{-3}, or $2N$ charges m^{-3} are housed in approximately $4R^3$. Thus, one frozen-in, single charge takes up a volume $1/(2N)$ and the distance between a pair is $R \sim 0 \cdot 7 N^{-1/3}$. However, owing to the finite electron temperature T_e, it is appropriate to operate here with a Debye cell of volume Λ^3 where Λ, the Debye radius (m), is as follows

$$\Lambda \simeq 7 \times 10^3 [T_e(\text{eV})/N_e]^{1/2} \tag{4.77}$$

The charge-density fluctuations in a Debye cell which contains $N_e \Lambda_3$ charges can be evaluated by equating the potential or field energy per unit volume to the electron kinetic energy per unit volume cell charge (or the electron partial pressure).

$$\tfrac{1}{2}\varepsilon_0 \overline{E^2} = N_e k T_e / N_e \Lambda^3 = k T_e / \Lambda^3 \tag{4.78}$$

For $N_e = 10^{21}$ m^{-3}, $T_e = 1$ eV $\sim 10^4$ K, $\Lambda \sim 2 \times 10^{-7}$ m and with $\varepsilon_0 = 8 \cdot 9 \times 10^{-12}$ F m^{-1} we have an average microfield of

$$\sqrt{(\overline{E^2})} \simeq 2 \times 10^6 \text{ V m}^{-1} \tag{4.79}$$

a value large enough to produce a reasonable Stark effect. The voltage across a cell is

$$E\Lambda \sim (2 \times 10^6)(2 \times 10^{-7}) = 0 \cdot 4 \text{ V} \tag{4.80}$$

With order 10^8 cells m^{-1} we have $\sqrt{10^8} = 10^4$ cells accelerating and the same number retarding per metre, so the change in energy is about 100 eV m^{-1}. The frequency of electron plasma oscillation and thus of the microfield is $f \sim 3 \times 10^{11}$ s^{-1} and the plasma wavelength is about 1 mm (equation 5.3).

Short-range and long-range interactions

From what has been said earlier it transpires that what are usually termed 'collisions', when the distances (impact parameters) are of atomic dimension, are better described as 'interactions', especially when the distances are large. In the case of a fully ionized gas or plasma it can be shown that short-range interactions, as between an electron and a nearby singly charged positive ion leading to a large change in direction, say by 90°, are less likely and few in number compared with multiple long-range interactions which occur between an electron and the many ions which, though relatively far away, lead in the end again to a 90° scatter. In the latter case the scattering is the result of a very large number of small interactions (pointing in all directions) which, for statistical reasons, finally give the large directional change. The choice of 90° is purely arbitrary, chosen for convenience to illustrate a large angle-scattering process.

Consider an electron of energy $\frac{1}{2}mv^2$ which is shot head-on against a singly charged negative ion of infinite mass. When the electron stops and is thus about to return (figure 4.27(a)) because the electrostatic potential energy, corresponding to that in a compressed spring, balances its initial kinetic energy, we have

$$\frac{1}{2}mv^2 = e^2/(d_c 4\pi\varepsilon_0) \tag{4.82}$$

which gives d_c, the distance of closest approach. If, however, the electron is shot with an impact parameter b towards the ion—the scatterer—then the deflecting force $f = e^2/b^2$, which persists for a time $t = 2b/v$. Thus, by balancing linear momentum we have

$$\Delta(ft) = \Delta(mv) = \frac{e^2}{b^2} \frac{2b}{v} = \frac{2e^2}{bv4\pi\varepsilon_0} \tag{4.83}$$

and for small scattering angles $\Delta\alpha$ we find, with equation 4.82 that

$$\Delta\alpha = \frac{\Delta(mv)}{mv} = \frac{2e^2}{mv^2 b} = \frac{d_c}{b} \tag{4.84}$$

This means that the deflecting angle in a single interaction is the smaller the larger the kinetic energy of the particle, i.e., the stiffer the particle 'beam' and the larger

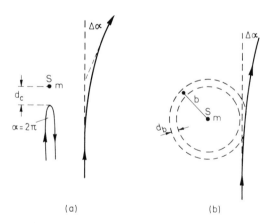

(a)   (b)

Figure 4.27. (a) Head-on electron collision with a single scattering centre S of infinite mass illustrating the distance d_c of closest approach.
Small impact parameter collisions result in small angle scattering.
(b) Plasma scattering.
Single centre S ($m = \infty$) is replaced by a large number N of cylindrical centres per cubic metre of radius b.

the distance of interaction expressed by the impact parameter. For a 100 eV electron interacting over $b = 10$ nm, 100 atomic radii, with an ion, such as H^+ or Hg^+, we find $\Delta\alpha \approx 1 \times 10^{-3}$ radian. The interaction time is about 3×10^{-5} s, i.e., of the order of one period of a wave of visible light.

To show under what circumstances the scattering in plasmas is predominantly caused by a series of small deflections, the following simplified argument may be helpful. The linear average of numerous randomly distributed angular deflections is, of course, zero. But the mean square of the deflections is a real finite number and, according to Einstein, corresponds to the mean square 'displacement' of a test particle on a 'random walk' in a fluid. Both types of displacement—in position and in angle—have been experimentally confirmed in fluids. Since from equation 4.84 $(\Delta\alpha)^2 = (d_c/b)^2$, it follows from figure 4.27(b), showing a cylindrical element of length l, radius b and thickness db (or volume $l2\pi b \, db$) containing N scattering centres per cubic metre, that with $C = d_c^2 \, l2\pi N$

$$\overline{\Lambda\alpha^2} = -C \int_{b=d_c}^{b=\Lambda} \frac{1}{b^2} b \, db = C \ln \Lambda/d_c \qquad (4.85)$$

where the limits are the Debye radius Λ, i.e., the distance between the point at which the field is screened by the charge cloud of opposite sign, and the scattering centre at d_c, the distance of closest approach of the projectile. It can be shown that for $1/e$ about 37 per cent of (multiple) deflections exceed $\Delta\alpha = \pi/2$, whereas for single scattering only a fraction of $\frac{1}{3} \ln (\Lambda/d_c)$ of all interactions lead to the same angle. Inserting numbers gives the result that only 4–8 per cent of all interactions are due to single large-angle scattering.

From the constant C we learn that $\overline{\Delta\alpha^2}$ is proportional to the concentration N of scatterers and the length l of the plasma cylinder, and inversely proportional to the square of the energy of the primary particle. Thus, single scattering predominates in 'thin' plasmas (as in Rutherford's thin foils), and multiple scattering in more extended plasmas, whereas in very dense, long plasmas the two types of scattering become equally likely.

It is of interest to discuss the transition from classical scattering in a feebly ionized medium to 'collective' scattering in a highly ionized plasma. A characteristic parameter may be the ratio ρ of the potential energy of two elementary charges at distance d and their kinetic energy, i.e., $\rho = e^2 N_e^{1/3}/4\pi\varepsilon_0 kT_e$. For $\rho \ll 1$, classical collisions predominate and collective scattering is negligible, and for $\rho \gg 1$ the reverse holds. This, however, turns out to be an unsatisfactory description.

Another approach is based on a comparison between the classical and the collective electron collision frequency ν_e and ν_{coll} respectively. The former is $\bar{v}_e/\lambda_e = \bar{v}_e q_e N_0$, q_e being the total electron–molecule momentum transfer cross-section and N_0 the gas number density; the latter is $\bar{v}_e N_i q_{e\text{-}i}$, where N_i is the ion number density and $q_{e\text{-}i}$ the electron–ion collision cross-section.

Thus, classical collisions fade out and distant encounters control the events, when $(q_{e\text{-}i}/q_e)(N_i/N_0) \gg 1$ or

$$\frac{N_i}{N_0} \frac{e^4}{2q_e (kT_e)^2} \ln\left[1 + \left(\frac{4\pi\varepsilon_0 kT_e}{e^2 N_i^{1/3}}\right)^2\right] \gg 1 \qquad (4.86)$$

and

$$q_{e\text{-}i} = \frac{e^4}{2(kT_e)^2} \ln\left[1 + \left(\frac{4\pi\varepsilon_0 kT_e}{e^2 N_i^{1/3}}\right)^2\right] \qquad (4.87)$$

Note that the ratio p appears here in a ln term. The mid-transition is reached when equation 4.86 is approximately 1 so that close and distant interactions occur at equal rates. For $kT_e = 3$ eV, $N_e = N_i = 10^{20}$ m^{-3}, the degree of ionization $N_i/N_0 = 1$ per cent and $q_e = 10^{-19}$ m^2.

The electric current in a fully ionized plasma

We have seen that in a moderately ionized gas far away from walls and electrodes the current density j is essentially determined by the drift motion of electrons, whereas the slow-moving positive ions hardly contribute to j but merely annul the negative space charge. Since according to equation 4.37

$$j \simeq \rho_e v_d = \rho_e \mu_e E = N_e (e^2/m)\tau E \qquad (4.88)$$

the 'conductivity' k_e, given by Ohm's law $j = k_e E$, at low E is

$$k_e = N_e \frac{e^2}{m} \tau \qquad (4.89)$$

where τ is the collision time $\tau_{e\text{-}m}$ between electrons and gas molecules, provided $N_e \ll N_0$. Under these conditions

$$\tau_{e\text{-}m} = 1/(N_0 \overline{q_e v_e}) = \overline{\lambda_e/v_e} \qquad (4.90)$$

In a fully ionized isothermal gas the collision time assumes another meaning and equation 4.90 fails. Firstly, because electrons and ions are the only constituents, $\tau_{e\text{-}i}(< \tau_{e\text{-}m})$ is the predominant time interval, so that μ_e, k_e and v_d are smaller than when neutrals are present; $\tau_{e\text{-}e}$ is usually negligible. Secondly, the presence of large numbers of charges requires that the frequent distant encounters must be included (Section 4.2), since only these are quantitatively important. If the ions were frozen-in, as in a lattice, a calculation of k_e in $(\Omega \text{ m})^{-1}$ with the plasma temperature T in K would yield, if $\tau_{e\text{-}e}$

$$k_e \sim 5 \cdot 5 \times 10^{-2} \frac{T^{3/2}}{Z \ln D_\Lambda} \qquad (4.91)$$

where Z is the ionic charge; $D_\Lambda = \Lambda/p = $ Debye length to mean impact parameter, a ratio of value of 10^2 to 10^3; and $\ln D_\Lambda$ is then of the order three to six, depending on N_e and T. The proportionality $k_e \propto T^{3/2}$ follows from equation 4.87, which shows that $q_{e\text{-}i} \propto \bar{\varepsilon}^{-2} \propto T^{-2}$, and by substituting this and $\tau_{e\text{-}i} = \lambda_{e\text{-}i}/v_e \propto 1/N_e q_{e\text{-}i} T^{1/2}$ in $k_e \propto N_e \tau_{e\text{-}i}$ in equation 4.89.

However, one of the most striking features is that the plasma electric conductivity k_e for completely ionized gases is independent of N_e. Formally this follows from equation 4.90, with $N_0 = N_e$, namely $\tau_{e\text{-}i} \propto 1/N_e$. Physically it means that since all electrons are taking part in the drift motion and their concentration does not depend on T (only their random motion), k_e cannot possibly depend on the electron number, except through Z.

If we take a hydrogen plasma ($Z = 1$) and $\ln D_\Lambda = 5$, the plasma conductivity k_e and resistivity $\rho = 1/k_e$ for various values of T and thus $\bar{\varepsilon} = kT$ (11 600 K = 1 eV) are as shown in table 4.5.

Large values of T, which apply to thermonuclear plasmas, give values of k_e and ρ which are comparable with or larger than those of metallic solid conductors at room temperature (for Cu at $T = 300$ K, $\rho \simeq 10^{-6}$ Ω cm).

There is another interesting effect connected with currents in plasmas. It has no parallel in solid conductors where, in general, the resistivity $\rho \propto T$ and is independent of the geometry, except for foils of thickness smaller than the de Broglie wavelength. In completely ionized plasmas, Ohm's law fails when the applied field E exceeds a certain value at which electron drift:random velocity > 1; in this case, a fraction of the drifting electrons in the energy distribution do not gain and lose the same amount of energy. The reason is that $\tau \propto 1/\overline{qv_e}$ varies with v_e, according to $q \propto 1/v_e^4$, and hence $\tau \propto v_e^3$. Those electrons which start a mean free path with a finite energy (or drift velocity) may gain along λ more than they lose in the next collision, and so become faster (runaway effect).

It can be shown that with $\Gamma = v_d/\bar{v}_e$ for a hydrogenous plasma, the 'runaway' effect will set in when $E >$ order 10^2–10^4 for $N_e/T_e \sim 10^{16}$

$$\tfrac{1}{2} m v_d^2 / kT_e > \frac{1}{\Gamma}; \qquad \Gamma = 6 \times 10^{12} \frac{T_e E}{N_e \ln D_\Lambda} \qquad (4.92)$$

(T_e in K, E in V m^{-1}, N_e in m^{-3}.) Thus, a fraction of the electron current increases with time (or distance) when E acts along a sufficiently long plasma conductor. This effect may also account for 'closed orbit acceleration' that yields very fast electrons (MeV), though the energy rise per orbit may only be of the order of 10 eV.

If we apply a field E to a fully ionized gas and a magnetic field B

Table 4.5. Plasma resistivity $\rho = f(T)$.

T(K)	$\bar{\varepsilon}$ (eV)	$k_e (\Omega \text{ cm})^{-1}$	$\rho (\Omega \text{ cm})$
10^4	1	50	2×10^{-2}
10^6	10^2	5×10^4	2×10^{-5}
10^8	10^4	5×10^7	2×10^{-8}

perpendicular to it, the electrons are forced to spiral around the magnetic lines with gyrofrequency ω (see Section 4.2). For strong fields, B, when the radius of gyration $\rho_1 < \lambda$ or when $\omega \gg \nu$, ν being the electron collision frequency, the plasma resistivity ρ_B (normal to B) can be several times the value given in equation 4.91. Such an increase in resistivity is to be expected, since any constraint in motion in the E-field direction should act like a scattering collision which reduces the 'mobility' of the electrons.

The thermal conductivity of a fully ionized gas

Consider a long, finite, vertical cylinder of a fully ionized plasma or of solid metal, electrically isolated, whose top end receives, for example, radiation which it absorbs, and whose bottom end is cooled in ice. In this way a constant temperature difference ΔT between the ends can be maintained and a heat flow F (in J s^{-1} cm^{-2}) downwards ensues, of magnitude

$$F = -c \frac{\Delta T}{\Delta l} \qquad (4.93)$$

where c is the thermal conductivity of the cylinder, provided no side losses occur, so the temperature gradient $\Delta T/\Delta l$ is constant.

The object of the following discussion is to express the heat flow and the associated phenomena in terms of concepts used in atomic physics. If the Wiedemann–Franz law holds, namely

$$c/k_e = \text{constant} \times T \qquad (4.94)$$

c and k_e being the thermal and electric conductivities respectively, which is true provided T is not too low and thermal conduction by vibrations or phonons in the solid is negligible, we are forced to conclude that both conduction phenomena are based on the same mechanism. Thus, if the flow of electric charge is due to directed motion of electrons, so must be the flow of heat. The model applicable to solids and gaseous plasmas is as follows.

The electrons (thought to form a gas) are kept at the top end at temperature T_1. Because of their larger random velocity they 'diffuse' downwards and deliver part of their energy to the ice which melts so that T_0 remains constant. Because the electron mass motion is accompanied by a motion of charge, we can associate it with a downwards flowing component of current. This means that the bottom end at T_0 receives an excess negative charge, making the T_1 end more positive. The corresponding electric field along Δl causes an electric current component flow upwards which is supplied by the 'colder electrons' at T_0. Since the system is isolated, $i_{\text{total}} = 0$ and $i\downarrow = i\uparrow$. The heat flow F is thus caused by electrons carrying kinetic energy

This picture gives the mechanism of the Kelvin effect and demonstrates that a current i passing through a wire forming an inverted V, whose ends are cooled, glows in that arm in which F and i are in the same direction and is cooler where F and i are in opposite directions (i corresponds to $i_{electron}$). Hence, the same i in the same conductor develops different amounts of heat, as if the resistances of the arms were different.

We saw in equation 4.91 that $k_e \propto T^{3/2}$ and with equation 4.94 we found $\tau \propto k_e T \propto T^{5/2}$. The thermal conductivity c (due to electrons), in cal K^{-1} cm^{-1} s^{-1}, is

$$c = 5 \times 10^{-12} \frac{T^{5/2}}{Z \ln D_\Lambda} \tag{4.95}$$

Again, in a strong field B the heat flow F perpendicular to B is reduced, but the thermal plasma conductivity c_B is now controlled by the positive ions because of their large radius of gyration.

electromagnetic, electron and ion waves: oscillations in plasmas

Consider a transverse or a longitudinal wave in an infinite medium. It may arise when a pulsed or a continuous perturbation is produced either at a discrete point or in a small volume. A wire connected to a signal source whose bare end is immersed in the ionized gas, or a beam of light or sound passing through a plasma or an electrolyte, are examples of media and perturbers. 'Probes' placed at different distances from the perturber, that measure for example the electric field, or light or sound signal intensity and the phase relative to the perturber, are used to determine the wavelength and speed of propagation, the rate of damping (attenuation of amplitude) and possibly the directional properties of the medium, as may be caused by external magnetic or other fields. The study of oscillations and wave motions in electric plasmas has proved to be of great value, such as in the field of communication, and has stimulated and facilitated research on account of many obvious analogies with modes of propagation in different media.

5.1. Electromagnetic and light waves interacting with a plasma

When a steady beam of visible monochromatic light of wavelength $\lambda = 400-700$ nm or frequency $f \sim 10^{15}$ Hz passes through a fully ionized plasma, the electric field E of the light beam interacts with the freely moving plasma electrons. Note that the beam direction is normal to E and that the electrons undergo forced oscillations in the E direction. For unpolarized light beams, E is vibrating radially in all directions in a plane normal to the beam. Since electrons moving in a general direction oscillate radially in this plane (figure 5.1), a certain amount of light is scattered sideways, and its intensity can be measured provided other scatterers or reflecting walls are absent. The scattered intensity is in general very small (see Section 5.2), but is directly proportional to the electron concentration N_e and hence is a useful means of measuring it.

If the electron random motion (v_e) is considered, it is seen in figure 5.1 that the light-scattering electrons have components of velocity towards and away from

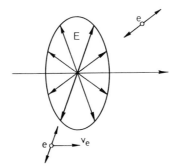

Figure 5.1. Unpolarized light wave and E field.
　Direction of propagation perpendicular to E. Electrons e, shown in 2 positions, forced to oscillate in E direction emit 'scattered' light; the lower one has an initial velocity v_e.

the observer. This gives rise to a Doppler broadening of the scattered spectral line, assuming monochromatic incident light. Hence the line width in the scattered light is a measure of $\bar{v}_e$ or the electron temperature T_e of the plasma. If $\Delta\lambda$ is the line width, $\Delta\lambda/\lambda = \bar{v}_e/c$, from which T_e can be found through $\frac{3}{2}kT_e = \frac{1}{2}m\bar{v}_e^2$.

In non-isothermal plasmas where $T_e \neq T_i$, the ion temperature T_i can be measured by observing Doppler broadening of spark lines that are emitted by excited positive ions, except for pure hydrogen plasmas.

For longer wavelengths, i.e., lower frequencies f, the intensity of the light scattered by free electrons is approximately proportional to $1/f^2$, whereas at higher values of f, for example for ultraviolet and X-rays, both true absorption (radiation damping) and Compton scattering occur (see Chapter 3). With beams in the centimetre and millimetre wavelength region, resonances and refraction effects are observed which depend only on the charge concentration in the plasma. This problem will be treated here and again in Chapter 12.

5.2. Electron plasma resonance frequency

Consider a small parallelepiped of length l parallel to x (figure 5.2) containing equal numbers of positive ions and electrons at zero temperature, the charges being smeared out (fluid) and uniformly distributed. Let us apply a weak perturbing pulse of electric field between its ends and determine the resulting motion of negative charge, assuming that the duration of the pulse is so short that the heavy ions remain in fixed positions. If then, on the left, a thin layer of thickness ξ with $d\xi/dx \ll 1$ is formed where the electron charge is temporarily in excess of the average (on the right there is an equal amount deficient), then from Poisson's equation the excess charge per unit volume is

$$e\,dN_e = eN_e\,\frac{d\xi}{dx} = \varepsilon_0\,\frac{d^2V}{dx^2}$$

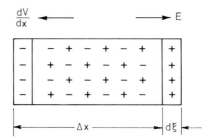

Figure 5.2. Perturbation of a plasma element of length Δx causing a transitory charge excess or deficiency at the ends.

An elementary charge e and mass m experience the force

$$-e\frac{dV}{dx} = \frac{e^2}{\varepsilon_0} N_e \xi$$

which, when equated to mass × acceleration, gives

$$m\frac{d^2\xi}{dt^2} + \frac{e^2}{\varepsilon_0} N_e \xi = 0 \tag{5.1}$$

the equation of a simple harmonic motion $\xi/\xi_0 = \cos(\omega t + \phi)$, ξ_0 being the amplitude and ϕ the phase between ξ and t. The angular frequency (in radian s^{-1}) is

$$\omega_e = \left(\frac{e^2}{m\varepsilon_0} N_e\right)^{1/2} \tag{5.2}$$

and the frequency (in Hz) of the oscillation is

$$f_e \simeq 9N_e^{1/2} \tag{5.3}$$

For moderately strong plasmas with $N_e = 10^{16}$ electrons m^{-3}, $f_e \simeq 10^9$ Hz corresponding to a wavelength of $\lambda \simeq 0.3$ m. For $N_e = 10^{20}$ m^{-3}, $f_e \simeq 10^{11}$ Hz and $\lambda \simeq 3$ mm. Hence, resonance, i.e., coherent electron amplitudes in the x direction, are to be expected when the 'applied' values of f or λ are of that magnitude. The amplitude itself can not, however, be derived from equation 5.1 unless the coupling between the free electrons and the other constituents of the plasma is known, which is seldom the case. It can be assumed that the amplitude is smaller than $N_e^{-1/3}$, i.e., the average distance between charges, here 10^{-5} to 10^{-7} m, which is smaller than the Debye length Λ. Obviously, f_e is independent of T_e, since a simple longitudinal concentration oscillation was assumed. If, therefore, an electromagnetic beam of variable f is sent through a plasma and its intensity at the exit observed, strong absorption occurs when $f = f_e$ and thus, from equation 5.3, N_e can be found. If, for example, N_e varies radially, as in the plasma of a tube, the maximum value of N_e is obtainable but not the local value.

I have assumed originally that both electrons and ions have zero random motion (T_e and $T_i \to 0$). The result for equation 5.3 thus holds only if $d\xi/dx$ is the same, independent of l (figure 5.2), i.e., if $l \to \infty$. In other words, propagation of the electron compressional (longitudinal) waves is not included in equation 5.1. In the following paragraph I shall introduce the dispersion problem by relating frequency ω with wavelength λ or wavenumber k of a wave.

You may notice that this section starts with a calculation of a resonance oscillation and ends with a note on progressive waves. It has to be realized that the production of oscillations requires an 'exciter' in the medium or coupled to it. Whether or not a detectable propagation of the disturbance through the medium results from it depends on the character of the disturbance, the size and boundaries of the medium (plasma) and other factors. If the 'exciter' produces an ordinary transverse electromagnetic wave (radio beam) of wavelength λ which traverses a plasma with a charge concentration N_e, corresponding to the electron plasma frequency ω_e in equation 5.2, then the actual angular electron frequency ω of the electron wave in an isothermal plasma is found to be

$$\omega^2 = \omega_e^2 + c^2 k^2 \tag{5.4}$$

where the wave number k is the number of cycles per unit length in the direction of propagation and c is the velocity of light, a result which will be discussed later. Also, since $\hbar\omega = h\nu = \varepsilon$ and $h\nu/c = \hbar k = p$, then the quantum energy $\varepsilon \propto \omega$ and its linear momentum $p \propto k = 2\pi/\lambda$, which gives the dispersion diagram $\omega = f(k)$ its physical meaning.

5.3. Ion plasma resonance frequency

The result obtained for electron oscillations (equation 5.2) suggests a similar form for the ion resonance frequency, if we write N_i instead of N_e (ω_i being the ion plasma frequency)

$$\omega_i = \left(\frac{e^2 N_i}{\varepsilon_0 m_i} \right)^{1/2} \tag{5.5}$$

It could be argued that now the ions oscillate in a sea of electrons of temperature T_e which closely 'follow' the heavy, slowly oscillating ions whose periods are longer by a factor $(m_i/m)^{1/2}$.

A general solution can be derived in a simplified form as follows. Taking the oscillations to be a simple harmonic motion, the force on a single ion (figure 5.2) caused by the space charge field is balanced by its inertial force

$$m_i \frac{d^2\xi}{dt^2} = -e \frac{dV}{dx} \tag{5.6}$$

where ξ is the ion's displacement, and V the small perturbing potential resulting from it, both given by a simple propagating plane wave

$$\xi = \xi_0 \exp\left[i(\omega t - kx)\right] \text{ and } V = V_0 \exp\left[i(\omega t - kx + C)\right] \quad (5.7)$$

any phase difference C between them being taken care of by a constant which vanishes by differentiation. The space charge equation for small perturbations in one co-ordinate is

$$\frac{d^2 V}{dx^2} = -\frac{1}{\varepsilon_0}\rho_{\text{net}} = \frac{e}{\varepsilon_0}(\Delta N_e - \Delta N_i) \quad (5.8)$$

where the electron component is given by the Boltzmann distribution, assumed to be established very quickly. Since $\Delta N_e \ll N_e$ because $eV/KT_e \ll 1$

$$\Delta N_e = N_e\{\exp\left[eV/KT_e\right] - 1\} \simeq N_e(eV/KT_e) \quad (5.9)$$

and again with $d\xi/dx \ll 1$

$$\Delta N_i = -N_e \frac{d\xi}{dx} \quad (5.10)$$

By substituting equations 5.9 and 5.10 in equation 5.8 we obtain with $a = 4\pi e N_e$

$$\frac{d^2 V}{dx^2} = a\left(\frac{d\xi}{dx} + \frac{eV}{kT}\right) \quad (5.11)$$

From equation 5.11 V can be eliminated by a short-cut: calculate the derivatives of ξ and V from equation 5.7 and substitute them in equation 5.11, yielding

$$-V[k^2 + (ae/kT_e)] = -iak\xi \quad (5.12)$$

By replacing $dV/dx = -ikV$ in equation 5.6, we have

$$\frac{d\xi}{dt^2} = ikeV/m_i = \frac{i^2 k^2 ae/m_i}{k^2 + (ae/kT_e)}\xi = -\frac{e^2 N_e/(m_i \varepsilon_0)}{1 + (e^2 N_e/kT_e k^2 \varepsilon_0)}\xi \quad (5.13)$$

which can be written as $d^2\xi/dx^2 = -\omega_i^2 \xi$. Hence the frequency of the wave is

$$\omega_i = 2\pi f_i = \left|\frac{e^2 N_e/(m_i \varepsilon_0)}{1 + [(e^2 N_e/\varepsilon_0 kT_e)(\lambda/2\pi)^2]}\right|^{1/2} \quad (5.14)$$

This shows that the ion plasma frequency given by equation 5.5 results when the denominator in equation 5.14 approaches unity, i.e., $k \to \infty$, $\lambda \to 0$, or when $\lambda \ll \Lambda_e$, $\Lambda_e \propto (T_e/N_e)^{1/2}$ being the Debye length (Section 5.4), which is of the order or smaller than 1 mm. Equation 5.15 has been confirmed experimentally in a plasma with singly charged ions at room temperature at large values of T_e. The other limit, holding for large values of λ (or N_e/T_e), yields the ion-acoustic wave of frequency

$$\omega_i = (kT_e/m_i\lambda^2)^{1/2} = k(kT_e/m_i)^{1/2} \tag{5.15}$$

Thus, whereas in the former case ω_i depends on N_e, in the latter it depends on T_e and k. In both cases, however, $\omega_i \propto (1/m_i)^{1/2}$. We conclude that for the lightest and heaviest atomic ions ω_i differs by more than one order of magnitude.

The wave or phase velocity of an infinite wave train is

$$v_p = f\lambda = \omega/k \tag{5.16}$$

From equation 5.16 the phase velocity of longitudinal (sound) waves of ions in a plasma for not too small values of λ is then

$$v_p = (kT_e/m_i)^{1/2} \tag{5.17}$$

Because $m_i \gg m$ the frequency f_i is at least 50 times lower than f_e, i.e., $f_i < 10^7$ Hz for $N_e = 10^{16}$ m^{-3}. However, whereas f_e was found to be approximately independent of λ, according to equation 5.14 f_i depends on λ; that is, f_i shows 'dispersion' is known in optics where the refractive index varies with the wavelength of light. The dispersion curve is here usually represented by the relation $\omega = f(k)$, $k = 2\pi/\lambda$ being the wave number in cm^{-1}. A gaseous medium without dispersion (curve 1) and with dispersion (curve 2) corresponding to equation 5.13 is shown in figure 5.3; curve 1 thus refers to a vacuum or a solid loss-free ideal dielectric.

In standard diagrams, the abscissa is the independent and the ordinate the dependent variable. Dispersion diagrams are exceptions: the quantity imposed on the system is the applied frequency (ordinate), the dependent variable the wavelength (abscissa). Yet to facilitate reading other texts, conventional dispersion curves were plotted. This means that, for example, when $\omega > \omega_i$ is applied to the system in figure 5.3, λ in the plasma cannot be measured because an ion wave does not develop.

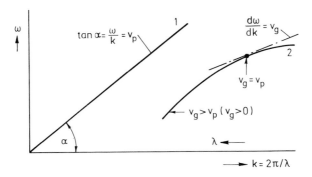

Figure 5.3. Dispersion in a plasma $\omega = f(k)$.
Curve 1 applies to an ideal dispersion-free dielectric; curve 2 to a real lossy medium; v_g and v_p are group and phase velocities; $\lambda = $ wavelength.

Also from figure 5.3 it is seen that the slope of curve 1 representing equation 5.16 and the phase velocity are related by

$$v_p = \tan \alpha = \omega/k = \text{constant} \qquad (5.18)$$

Since the group velocity v_g (associated with the propagation of the power of the perturbation) is defined by $v_g = d\omega/dk$, when equation 5.18 holds, $v_g = v_p$ for any real value of k.

In a 'curve 2 medium', however, $\tan \alpha$ and v_p vary with k. It is seen that as k is raised $\tan \alpha$ goes through a maximum. Hence the group velocity v_g is first $> v_p$, at α_{max} $v_g = v_p$ and later $v_g < v_p$. Also here $v_g > 0$ throughout.

Of special interest is figure 5.4, the dispersion curve of a fully ionized (atomic) hydrogen plasma, illustrating equation 5.14. Since such a plasma can only be approximately realized, no fully satisfactory tests of the theory of the ion space charge (or ion sound) wave have been made in H_2. For convenience a double-logarithmic plot is used. In the lower k range $\omega_i \propto k$ and hence the phase velocity equals the constant velocity v_s of the ion sound wave. At higher k values ω_i approaches asymptotically an upper limit ω_{pi}, the cut-off frequency, which means that $v_g = d\omega_i/dk \to 0$; no ion wave propagation materializes when $\omega > 10^9$ rad s^{-1}, although at higher ω_i the applied electric field continues to produce plasma oscillations.

So far ion waves progressing without loss in amplitude, or losses assumed to be negligibly small, have been discussed, as can be seen from equation 5.13. Kinetic theory predicts that losses are likely to be considerable when T_i

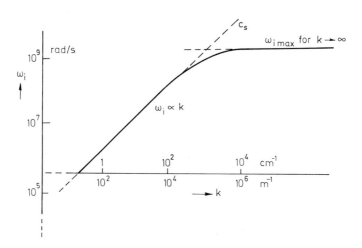

Figure 5.4. Calculated dispersion curve of a fully ionized H plasma (angular frequency ω vs. wave number k).
$k = 2\pi/\lambda$; $T_i = 10^3$ K; $T_e = 10^5$ K; $N_i = 10^{18}$ m^{-3}; ion space charge (ion sound) velocity $v_s = 5 \times 10^4$ m s^{-1}.

approaches T_e, because the ion random velocity $\bar{v}_i$ approaches the ion acoustic wave velocity v_s and so-called 'collisionless damping' increases. Besides that, collisional losses may occur. It has to be noted that collisions create discontinuous disturbances whereas electrostatic collective effects engaged in energy transfer from electrons trapped by potential waves are continuous.

The collisionless abstraction or redistribution of wave energy is the so-called 'Landau damping'. This is caused by particles moving with almost phase velocity v_p which suffer from large changes in their trajectories. Particles with $v < v_p$ are accelerated and extract energy from the wave, whereas particles with $v > v_p$ are decelerated. Yet the net energy change only causes a dampening if the first derivative of the bulk velocity distribution df/dv is negative for $v = v_p$. The dampening is thus due to there being more slow ones accelerated than vice versa. The overall result is an exponential damping of the wave amplitude. It follows that a damping factor has to be included in equation 5.7 which—as in optics—leads to a complex wave number k. As usual, the real part of k is the same as the modulus, provided the losses are reasonably small (a result as found in our earlier equations), whereas the imaginary part gives expressions identical with those for a dielectric with losses. In this way the dispersion of gaseous plasmas is seen to be formally closely related to solid 'lossy' insulating substances.

5.4. Electron waves

We now return to the propagation of longitudinal electron (compression) waves in plasmas and to equation 5.4. It can be shown that v_{pe}, the electron velocity, depends on the applied frequency ω and the electron temperature T_e according to

$$v_{pe} = \left[\frac{3kT_e}{m_e} \frac{1}{1 - (\omega_{pe}/\omega)^2} \right]^{1/2} \tag{5.19}$$

when ω approaches ω_{pe} and thus $v_{pe} \gg \bar{v}_e$ the (root mean square) random electron velocity given by $\bar{v}_e = (3kT_e/m_e)^{1/2}$. There is good experimental verification of this mode of propagation. Since $v_{pe} = \omega/k$ the dispersion $\omega = f(k)$ can be easily obtained; it is plotted in figure 5.5.

5.5. Electromagnetic waves of low and high intensity

The interaction between light or radio waves and a weak plasma is discussed in Chapter 12, where the ionosphere is considered. Therefore, it suffices to state that the electromagnetic wave velocity $v_{e.m.}$ in a plasma, defined by T_e and ω_p,

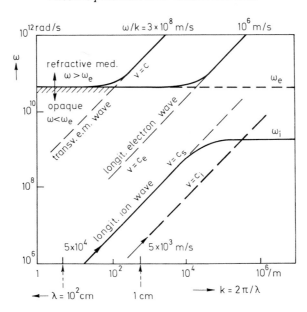

Figure 5.5. Dispersion curves of the H plasma (figure 5.4) which include electromagnetic, electron and ion waves for $B = 0$.
c = velocity of light = $(1/\varepsilon_0\mu_0)^{1/2} = 3 \times 10^8$ m s^{-1}; c_e = electron acoustic (compression) wave velocity = $(\gamma_e kT_e/m_e)^{1/2}$; c_s = space charge (electron acoustic) wave velocity = $(\gamma_s[kT_e + kT_i]/m_e + m_i)^{1/2}$; c_i = positive ion acoustic wave velocity = $(\gamma_i kT_i/m_i)^{1/2}$; $\gamma = c_p/c_v = (n+2)/n$; n = degrees of freedom. For electron wave motion in one dimension $n = 1$, $\gamma = 3$ and $v_e = (3kT_e/m_e)^{1/2}$, see equation 5.18. For rare gases $n = 3$, $\gamma = 5/3$; for air at $T = 3000$ K (full rotational levels) $n = 3 + 2$, $\gamma = 7/5$.

can be obtained by replacing $\bar{v}_e$ by c in equation 5.19. Hence, for small losses we have

$$v_{e.m.} = \left[\frac{c}{1 - (\omega_p/\omega)^2} \right]^{1/2} \tag{5.20}$$

Putting $v_{e.m.} = \omega/k$ gives the dispersion equation 5.4 again

$$\omega^2 = \omega_e^2 + (ck)^2 \tag{5.20a}$$

Figure 5.5 illustrates the dispersion curves of the plasma of figure 5.4 when all three types of wave are considered. For the waves in equation 5.19 or 5.4 we find for low k values $\omega \simeq \omega_e$ = constant and for high k values $\omega_e \simeq ck$ as shown in the left top corner. The longitudinal electron and ion waves occur for $\omega > \omega_e$ and $\omega < \omega_i$ only.

Before we leave this subject it might be of interest to note that in recent studies of wave propagation, instead of small perturbations (amplitudes of electric field, space charge, etc.), large disturbances are activated. As a result the starting equations become non-linear and the analysis, in general, is more involved. In spite of this, the appearance of non-linear waves in future thermonuclear plasmas demands an intimate knowledge of these effects which may impose restrictions on the design of the enclosure.

5.6. Plasma waves in magnetic fields

A great variety of other types of waves are produced when an electron ion gas is subject to a constant magnetic field B. Now a disturbance of frequency ω will be considered when $\omega < \omega_{ci}$, where ω_{ci} is the ion–cyclotron frequency.

For a charge Ze of mass M the 'Larmor' (cyclotron) frequency (Chapter 4)

$$\omega_{ci} = \frac{Ze}{M} B \tag{5.21}$$

and the corresponding radius of the path (precession)

$$\rho = Mv_\perp / ZeB \tag{5.22}$$

$v_\perp$ being the velocity component normal to B. To keep $\omega \ll \omega_{ci}$, for given M, B must be large; that makes ω_c large, and ρ becomes very small. It appears then as if the charges are fixed to the magnetic field lines like balls to wires pierced through their centres. A perturbation of the plasma normal to B (acting on the elastic springs, alias magnetic lines, uniformly loaded with masses M) will progress as a transverse wave in the field B, since ions, plasma and field are virtually sealed together. Since the longitudinal tension per unit area in the field is B^2/μ_0 and the ion mass per unit length and unit area is MN_i, the partial differential equation of the elastic string (figure 5.6) reads, with x, the direction of the unloaded string, and y, the direction perpendicular to it

$$\frac{\delta^2 y}{dx^2} = \left[\frac{MN_i}{B^2/\mu_0} \right] \frac{\delta^2 y}{\delta t^2} \tag{5.23}$$

Its derivation is found in text books on mechanics; $\mu_0 = 4\pi \times 10^{-7}$ H m^{-1}.

In equation 5.22 the dimension of the bracketed expression is the inverse of a (velocity)^2. This is the velocity of propagation of the disturbance—the Alfven ion velocity

$$v_{Ai} = [(B^2/\mu_0)/MN_i]^{1/2} \tag{5.24}$$

For protons, $B = 10^{-4}$ T (1 gauss), $M = 1 \cdot 67 \times 10^{-27}$ kg, $N_i = 1 \times 10^{11}$ ions m^{-3}, $v_{Ai} = 6 \cdot 9 \times 10^6$ m s^{-1}. The electron contribution is negligible since $m_e \ll M$. The result holds for disturbances of $\omega \ll 10^8$ s^{-1}. Equation 5.24 can also be obtained by converting half of the magnetic field energy into the energy of the transverse ion wave. When $\omega \ll \omega_{ci}$, circularly polarized waves (anti-clockwise) appear whereby ions and field lines rotate in the same sense.

The value of p increases with ω and would rise to ∞ when $\omega \to \omega_{ci}$, if damping were absent. When $\omega > \omega_{ci}$ (clockwise) rotation sets in and p decreases as ω rises, because now inertial effects become less important than elastic effects, until finally the field lines rotate about the stationary ions. For such 'hydromagnetic' waves, B represents the sea, and the water droplets the distributed electric charges.

In the presence of a field B many different types of wave can appear in the plasma, such as cyclotron waves and 'Bernstein modes', but a discussion of their properties would by far exceed this elementary introduction. Advanced treatises are listed in Further Reading. Finally, a note on the question of quantization of waves has to be added. In analogy to 'light quanta' which replace electromagnetic waves, or 'phonons' replacing lattice vibrations in solids, 'plasmons' of energy $\hbar\omega$ have been introduced as the 'quanta of plasma waves'. In this way the interaction between a particle and a wave can be described as a collision between a particle and a plasmon.

CHAPTER 6

nature and properties of plasmas

6.1. Introduction

If we were to fill an imaginary box with a partially or fully ionized gas which had been produced by separating electrons from neutral atoms, and if all particles were uniformly distributed throughout the box, an observer outside this enclosure (assumed to have no walls) would find no electric field since such a 'neutral plasma' contains equal numbers of positive and negative charges. If the observer, however, were to move sufficiently near to the boundary of the enclosure, say to distances of the order of the average distance between any two particles, or were to enter the box, he would observe very strong electric fields indeed which, because of the motion of the particles, vary in magnitude and direction. These fluctuating electric fields—the micro-fields—are a prominent feature of all gaseous matter in a highly ionized state. They will be discussed later.

When a gas is being ionized the number of particles increases, assuming a constant mass of gas. At large degrees of ionization the total pressure P can rise considerably above the original neutral gas pressure; this has been confirmed experimentally. Dalton's law thus apparently holds, as with a mixture of gases

$$P = p_{gas} + p_{ion} + p_{electron}$$

where the ps are the partial pressures of the three components.

Crookes (1879) was probably the first to describe the plasma state. He called it 'the fourth state of matter'. The term plasma, however, was first used by Langmuir (1929) in his investigations of a long uniform positive column of an electric discharge 'moulding' ordinary neutral, excited and charged particles and light quanta which form the ionized gas. A mathematical definition of a plasma was given in Chapter 4.

A plasma is often the result of the action of an electric discharge in a gas. To produce it, energy is taken from a source which maintains a potential difference across the gas and the electric field drives a current through the gas. The steady state, where energy losses balance the energy input, is fundamentally different from the thermal equilibrium which lends itself more easily to theoretical treatment; in practice it is difficult to realize because thermodynamic systems with zero losses can rarely be attained. In this ideal case the motion, interaction and transfer of particles from one energy state into another (excited atoms) is uniquely described by the temperature of the system.

If the system is in a thermodynamic equilibrium, then, according to a principle first enunciated by Klein and Rosseland, for every process going in one direction there must be a similar one going in the reverse direction, thus keeping the populations of particles in the various energy states the same, apart from fluctuations. For example, per unit time the number of atoms in the ground state which by absorbing light quanta are raised to a certain higher energy state is the same as the number of transitions from this energy state down to the ground state (accompanied by emission of quanta of that energy). Or, the number of atoms hit by electrons and producing positive ions and secondary electrons is the same as the number of ions recombining with free electrons—the recombination energy coming from the kinetic energy of the electron which happens to be near enough to neutralize the positive ion.

In the steady state the plasma populations of ions, electrons, excited particles and quanta similarly do not vary with time; but here energy is continuously fed into and lost by the system. For this reason the temperature is not, strictly speaking, a rigorous parameter fit for a description of such a plasma as a whole; though it is often convenient to apply it to the various populations. This is particularly so if the energy exchange between different groups of particles is restricted, as for example between gas molecules and electrons. Because of the large ion/electron mass ratio of 2000 to 500 000, elastic collisions can result in the formation of an 'electron gas'; the energy distribution of the electrons is thus often described by an 'electron temperature'—equivalent to the average energy of a particle in terms of kT. Similarly, 'ion temperatures' and 'excitation temperatures' have been used to describe the corresponding distribution.

However, for many problems it is necessary to know the type of distribution, which is not always Maxwellian. For example, 'very hot' transient plasmas often contain a considerable number of fast ('runaway') electrons as a result of the electric field acting on the electrons which travel through a tenuous gas whose collision cross-section decreases with increasing energy.

6.2. Production of a plasma

We shall distinguish between arrangements which give a steady plasma and those which give transient plasmas. An example of a steady plasma is shown in figure 6.1. An arc discharge in an air stream or any other gas at atmospheric or at lower pressure is set up between two carbon or water-cooled metal electrodes. Through the hole in the anode a hot plasma jet of considerable length (1–10 cm) escapes which represents a neutral plasma. The net current flowing through the

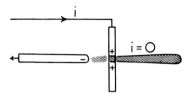

Figure 6.1. Quasi-neutral continuous or discontinuous plasma jet at high gas pressure.
Moderate but equal gas and electron temperatures. Zero net current *i* in plasma jet.

jet is zero and so are the macroscopic electric and magnetic fields in and around it. The arc currents are usually between 10 and 1000 A; gas temperatures up to 20 000 K can be obtained with charge densities up to 10^{16} cm^{-3}.

Figure 6.2 shows a steady uniform positive column of a glow or arc discharge in molecular or rare gases at either low or high pressure. It is distinct from the plasma in figure 6.1 in that it carries a current and thus there are axial and radial components of the electric field and a tangential component of the magnetic field. There are many variants of this arrangement; for example, cooled constrictions are used in order to obtain high current densities along a short length of the current path, which results in higher electron densities and gas temperatures.

Attempts are being made at present to study steady fully ionized 'synthetic plasmas' which are essentially free of neutral gas. Figure 6.3(*a*) shows that this could be obtained by emitting electrons from one electrode and positive ions from another electrode, so that the corresponding beams of charges move in opposite directions. If the number density of charges in the two beams can be made sufficiently large and the neutral density sufficiently low, collisions between the charges will be predominant and a fully ionized gas plasma should form. The same object can be achieved by emitting the two beams with opposite charges from the same end of a tube (figure 6.3(*b*)); the ions can be formed, for example, by surface ionization. (Further Reading: Carré *et al.*, 1981, Chapter 7.)

A fairly simple way of producing transient travelling plasmas is by means of shock tubes. Figure 6.4(*a*) shows an apparatus which consists of a section in which a transient gas discharge of high intensity is produced by discharging a condenser through rare or molecular gases at low pressure. The adjoining shock tube which contains the same gas at the same initial pressure carries the shock wave with the associated plasma down the tube where its properties can be investigated. Another method (figure 6.4(*b*)) of producing shock wave and plasma

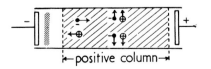

Figure 6.2. Steady glow-discharge plasma at low gas pressure.
Gas at room temperature, electrons at high temperature. Finite current *i* in the plasma.

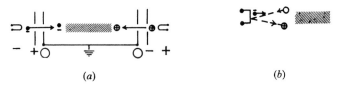

Figure 6.3. Steady 'synthetic' or 'quiescent' plasma. Gas is practically absent, and hence, if the interaction between the charges is strong, a fully ionized plasma forms.
(*a*) Electrons and ions produced separately, and move in opposite directions.
(*b*) Electron-emitting surface acts as positive ion emitter when struck by neutral atoms (e.g., W by Cs), so both charges move in the same direction.

is by exploding electrically a mixture of hydrogen and oxygen in a combustion chamber separated by a thin breakable diaphragm from the shock tube, which is evacuated or contains a suitable gas. Both of these methods project a strongly luminous and very hot plasma along the tube.

A toroidal plasma (of vortex ring or 'smoke ring' shape) can be produced by using a tube containing what is really a special type of spark plug (figure 6.5). Finally, a metal vapour plasma may be produced by passing large currents through thin fuse wires (figure 6.6); melting occurs preferentially in certain sections of the wire (as pictures taken with high-speed cameras show), so that the plasma is a series of disconnected blobs whose investigation presents considerable difficulties.

6.3. General physical properties of the plasma

A plasma (figure 6.7), although of course gaseous, has certain similarities with matter in non-gaseous forms. If the charges, instead of moving at random, were frozen and assembled in a regular pattern, the similarity between such a 'low-temperature plasma' and a crystal would be obvious. If, however, the charges consisted of positive and negaative ions (instead of electrons), a comparison of this 'ionic plasma' with an electrolyte, or with ionic crystals, would be appropriate. Moreover, certain properties of semi-conductors can be

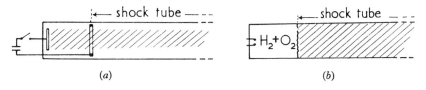

Figure 6.4. Plasma pulses produced in shock tubes.
Zero net current in plasma, gas pressure low or moderate or vacuum.
(*a*) Combustion produced by intense gas discharge; the initial gas pressures in the discharge tube and shock tube are equal.
(*b*) Ionization, excitation and pressure waves produced in the combustion chamber separated by a diaphragm from the shock tube section.

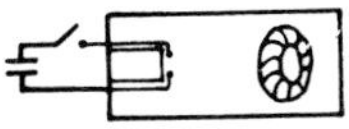

Figure 6.5. Plasmoids, which are vortex rings originating in the adsorbed gas released from the insulator carrying the spark gap. A current pulse is observed along the circular axis of the torus.

compared with ionized gases containing 'impurity atoms' which give the plasma special properties.

How closely a plasma resembles a crystal can be expressed by the ratio C of the potential energy of a plasma charge e separated d cm from its closest like-neighbour (the crystal energy), and the kinetic (random) energy of the electrons. For a cube, $d = N^{-1/3}$, where N is the number of charges per cm^3

$$C = (e^2/d)/kT = e^2 N^{1/3}/kT \tag{6.1}$$

T is the temperature of the 'electron gas' in K. In general, $C \ll 1$ and so the crystal-like behaviour of the plasma cannot be expected to be very pronounced. A comparison with electrolytes, however, seems to be more fruitful.

It can be shown that there exists in the plasma a characteristic distance, corresponding to the lattice parameter d, which describes the size of a 'plasma cell'. Let us assume that the cell size is large compared with the diameter of the ion so that we can suppose the plasma charges to be distributed uniformly. Consider a positive charge in such a plasma. It will attract electrons and thereby increase the electric field in its neighbourhood; it will slightly repel the heavy positive charges, which reduces the field at greater distances from it. Further re-arrangement of the other surrounding charges will lead to a still stronger field decrease, certainly more rapidly than the inverse square of the distance. Roughly speaking the field is confined to a finite region; outside this cell the field is essentially zero. This picture can be exploited quantitatively. In order to find the approximate range over which a field E exists we move all electrons from a region D of the uniformly fully ionized plasma (figure 6.8) to the right of the vertical line a. On the left of this boundary there is now a sheath of positive charges of constant density N. In this case Poisson's equation reads for $x > 0$

$$\frac{d^2 V}{dx^2} = 4\pi eN \tag{6.2}$$

and by integrating it twice we obtain (with the boundary conditions $E = 0$ at $x = D$ and $V = 0$ at $x = 0$)

$$E = -4\pi eN(D - x) \tag{6.3}$$

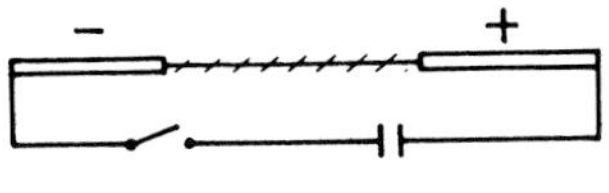

Figure 6.6. Vapour plasma obtained by fast fusing a thin metal wire.

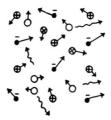

Figure 6.7. Section of a plasma showing positive ions, electrons, excited and normal atoms and photons.

and

$$V = 2\pi e N (2Dx - x^2) \qquad (6.4)$$

Thus the work W to move an electron through the space charge from $x=0$ to $x=D$ is $W = 2\pi e^2 ND^2$. If this is done by the thermal energy of the electron gas it can be equated to the kinetic energy of an electron in one direction, $\frac{1}{2}kT$. Therefore the 'shielding distance D' (in cm), at which the field has reached zero value, is

$$D = (kT/4\pi e^2 N)^{1/2} \simeq 7(T/N)^{1/2} \qquad (6.5)$$

In ordinary discharges with $T = 10^4$ to 10^5 K and N between 10^9 and 10^{14} cm^{-3}, D is between 10^{-1} and 10^{-4} cm. Thus the shielding distance is usually larger than the average distance ($N^{-1/3}$ between charges which is 10^{-3} to 10^{-5} cm. This is also true for partially ionized or non-isothermal plasmas; the differences are insignificant. If we had assumed spherical symmetry and a Boltzmann distribution of charge density [$N \propto \exp(-eV/kT)$], the calculation would have given a distance called a 'Debye radius' whose value is about the same as D from equation 6.5.

Equation 6.5 shows that the larger N is, the stronger is the field necessary to separate the charges and the smaller the space D to achieve this. As T is raised, the random velocity of electrons rises; this favours diffusion and separation, and the distance D, over which now the weaker field extends, is increased.

When a plasma is bounded by an insulating wall or probe so that an equal number of charges of both signs are forced to arrive and combine to give zero current to the probe, the electrons keep the wall at a potential which is negative

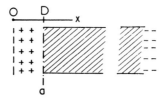

Figure 6.8. Fully ionized plasma (shaded) and Debye's shielding distance D.
$x =$ axial co-ordinate.

with respect to the plasma. Thus only the fast electrons will reach the wall together with an equal number of positive ions, while the slow electrons are repelled. The wall field extends over a distance ('the sheath') which is of the order D.

It is apparent that the shielding plays a dominant role in plasmas. For this reason a plasma is often defined as an ionized gas in which $D \ll R$, R being the tube radius.

Although the linearly averaged electric field X at a point in the plasma is zero, the averaged modulus $|X|$ is of the order $e/(d/2)^2 \simeq 4eN^{2/3}$ and in fact is probably three to five times larger. For $N = 10^{12}$ cm^{-3}, $|X| \simeq 200$ V cm^{-1}, and for larger N the microfield produces an 'internal Stark effect' in the line emission spectrum. The atoms in the plasma 'see' a field X which varies with a frequency (noise) corresponding to the velocity of an electron crossing d; in our example it is $\sim 10^{10}$ Hz.

Some of the physical properties of a fully ionized plasma will now be derived from first principles.

6.4. Dielectric properties

Since a steady uniform electric field will convey the charges in a plasma to the respective electrodes so that a direct current circulates in the circuit, there is no d.c. dielectric constant of plasma condenser. The a.c. dielectric constant, the corresponding capacitance of a plasma and its dependence on the frequency of the field, however, can be understood.

When an electric field of amplitude X_0 and angular frequency ω acts upon a charge e of constant mass m *in vacuo*, the force is $eX_0 \sin \omega t = m(dv/dt)$, its velocity $-v = (eX_0/m\omega) \cos \omega t$ and the current density due to the oscillating motions of N charges per cm^3 is

$$j_m = -eNv = -(e^2 N X_0/m\omega) \cos \omega t \tag{6.6}$$

showing that j_m lags by $\pi/2$ behind X and is thus an inductive component. The other component j_c is due to the surface charges on the condenser plates and is of magnitude

$$j_c = \frac{\varepsilon_0}{4\pi} \omega X_0 \cos \omega t \tag{6.7}$$

Since $j = j_c + j_m$, we find that with $j/j_c = \varepsilon/\varepsilon_0$ and $\varepsilon_0 = 1$ (vacuum)

$$\varepsilon = 1 - \frac{4\pi e^2 N}{m\omega^2} = 1 - \left(\frac{\omega_p}{\omega}\right)^2 \tag{6.8}$$

i.e., the effective dielectric constant is reduced by charges oscillating in an evacuated condenser. This is to be expected because the inertial effects in mechanics are analogous to inductive effects in electric circuits. Thus, provided N is not too large, and changes of the local field X by space charges are small where

$$X = X_{\text{applied}} - f(N)$$

and furthermore, if the field X is set up either by electromagnetic waves or by electrodes which are so far away that charges cannot reach them, conditions can be such that the dielectric constant in equation 6.8 can become zero or even negative. The former is the case when a vertically emitted electromagnetic wave reaches a height where the net electron concentration in the ionosphere is so great that it is reflected back to earth. This happens for frequencies of order one megahertz at a concentration of about 10^5 cm^{-3}.

If, however, instead of a vacuum we have a gas plasma, then the oscillating charges collide with gas molecules, are scattered, and acquire a random velocity. Hence their synchronous motion which originally lagged behind the field is now interrupted. After a collision a charge starts in general with an 'in phase' component of velocity. Thus work is done by the field on the charges. Their energy is transferred by collisions to the gas, whose temperature rises slightly. When charges are acted upon by an a.c. field X and their drift velocity v_d (the component parallel to X) is retarded by a force $c_f v_a$, where c_f is a constant 'coefficient of friction', then the balance of forces gives

$$m(dv_d/dt) + c_f v_d = eX_0 \sin \omega t \qquad (6.9)$$

Introducing the mobility μ, we have $c_f \mu X = eX$ and assuming (Langevin) $\mu = (e/m)(1/v)$, v being the collision frequency for a charge colliding with gas molecules, then

$$c_f = (e/\mu) = vm \qquad (6.10)$$

With this result the solution of equation 6.9 is

$$v_d = \frac{eX_0 \sin (\omega t - \phi)}{m (v^2 + \omega^2)^{1/2}} \qquad (6.11)$$

with $\tan \phi = \omega/v$, and the current density is

$$j = \{e^2 NX_0 \sin (\omega t - \phi)\}/m(v^2 + \omega^2)^{1/2} \qquad (6.12)$$

In general, j lags behind X by a phase angle $\phi \leq \pi/2$. For $\omega \to 0$ the motion is simply controlled by the ion mobility. The equation for j contains a 'complex conductivity; $\sigma = e^2 N/m(v^2 + \omega^2)^{1/2}$; this implies a complex dielectric constant whose real part is

$$\varepsilon = 1 - \frac{4\pi e^2 N}{m(v^2 + \omega^2)} = 1 - [\omega_p^2/(v^2 + \omega^2)] \qquad (6.13)$$

We conclude that collisions reduce the effective value of the dielectric constant. It has to be pointed out that equation 6.11 is an approximation which holds strictly only for ions in low fields; it can be used for electrons in those molecular gases where deviations from a constant mobility are not too large.

In plasmas of finite size the use of ε is often of doubtful value since it contains no real dipoles. Thus to apply ε the amplitude of oscillation must be smaller than D or the size of the vessel.

An interesting application of the dielectric properties of the plasma is the 'hydro-magnetic capacitor'. This is essentially a small concentric cylindrical condenser filled with a gas at low pressure in which a dense plasma is established for a short time. By means of a magnetic field normal to the electric field the plasma is made to rotate so that it represents a gaseous flywheel of small inertia (figure 6.9).

The dielectric constant of the linear system is easily calculated. Let the particle density ρ and velocity v be uniform throughout the gap. Then, neglecting losses and equating the total field energy to the sum of electrostatic and kinetic energies (all per unit volume) we have

$$\varepsilon \frac{X^2}{8\pi} = \varepsilon_0 \frac{X^2}{8\pi} + \tfrac{1}{2}\rho v^2 \qquad (6.14)$$

Since the velocity v is produced by a magnetic field B normal to X

$$v = Xc/B \qquad (X/B \text{ in S.I.}) \qquad (6.15)$$

where c is the velocity of light and X and B are expressed in e.s.u. and e.m.u., respectively. Note that the direction of v is independent of the sign of the charge and of the mass. Eliminating v, we obtain for the effective dielectric constant

$$\varepsilon = 1 + 4\pi\rho c^2/B^2 \qquad (6.16)$$

With argon at a pressure of $0 \cdot 1$ mm Hg, $\rho = 2 \cdot 3 \times 10^{-7}$ g cm^{-3}. For $B = 10^3$ G $= 10^{-1}$ T, $\varepsilon = 2 \cdot 5 \times 10^7$. With $X = 3$ e.s.u. $= 0 \cdot 9$ kV cm^{-1} the energy per unit volume from equation 6.14 is 10 J cm^{-3}. However, much higher energy densities have been obtained. A particular feature of this condenser is its low inductance; hence a very short discharge time ($\simeq 10^{-5}$ s), depending mainly on the length of the leads in the consuming system, can be expected. For a 'vacuum capacity' of 1 pF the dynamic capacity may be $\simeq 100\ \mu$F. When the plasma velocity is high ($>10^6$ cm s^{-1}) as well as its density, the centrifugal plasma

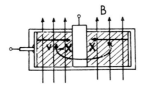

Figure 6.9. Rotating plasma capacitor.

pressure blows the magnetic field slightly outwards and ε rises by another factor of about two.

6.5. Optical properties

Though the relation between dielectric and optical properties of gases has been known for a long time, their study has only quite recently been extended to plasmas. Thus from measurements of the refractivity and dispersion of a plasma, the degree of dissociation and ionization can be determined.

The refractive index of an ordinary gas in excess of that of vacuum $(n-1)$ is equal to the specific refractivity times the gas density ρ, provided the density is not too high; for a gas mixture it is equal to the sum of the products of partial refractivities and densities. Again, for a single gas the excess refractive index $n-1$ is proportional to the number concentration of particles N, and the mean polarizability α, which in turn depends on the oscillator strength for the transition in question, the frequency at which absorption occurs and the frequency of the light. This applies only to a gas whose particles have no permanent, but only an induced dipole moment. In a gas mixture, or in excited or ionized gases, the optical properties are assumed to be additive in character.

In general it is difficult to calculate the polarizability of a plasma, since simple expressions fail to describe α near absorption frequencies. In all cases where α is known, n can be calculated with ϕ_i (see figure 2.5)

$$n - 1 \simeq 2\pi \sum_i \alpha_i N_i \qquad (6.17)$$

where N_i is the number of molecules per cm^3 of the ith species. On the other hand, the total refractive index can be determined interferometrically, and by using light of different wavelength and knowing their individual dispersions the fractions of the various components in the gas can be derived. If excited molecules are present, then the corresponding values α^* are to be used in equation 6.17 which are usually considerably larger than those of molecules in the ground state.

The refractivity of an ionized gas can be investigated similarly. Electrons make a major contribution to it. The result can be inferred from equation 6.8 with $\omega_p^2 = 4\pi e^2 N/m$. This gives with zero damping

$$n - 1 \simeq -\omega_p^2/2\omega^2 \qquad (6.18)$$

and describes the dispersion where ω is the angular frequency of light and ω_p the angular plasma frequency. Since the latter is proportional $N^{1/2}$, the electron concentration and the degree of ionization can be derived from interferometric measurements. It is unnecessary to know the absolute value of n if measurements of n_1, n_2 at two different wavelengths λ_1, λ_2 are carried out. Thus N follows from

$$n_1 - n_2 \simeq -A(\lambda_1^2 - \lambda_2^2)N \tag{6.19}$$

where A is a known numerical constant.

Such measurements have been made on shock waves in argon at pressures of a few mm Hg, with interference filters and visible light beams normal to the shock. Electron concentrations of 10^{16} to 10^{17} electrons per cm³ have been derived from interferograms. It is important to remember that this is the favourable range of application. Moreover, it is possible to derive in the same way the relaxation times, i.e., the time dependence of the degree of ionization.

6.6. Electric conduction

A plasma of infinite dimensions maintained at a given temperature and charge concentration possesses an electric conductivity σ which is determined by collisions between electrons and gas molecules

$$\sigma = j/X = (eNv_d)/X = eN\mu \tag{6.20}$$

provided the electrons carry the current and the contribution by positive ions is negligible. Here the symbol μ stands for mobility. In general the mobility of electrons μ^- is a fictitious quantity since it depends strictly on X and N. However, in many cases equation 6.20 gives the right magnitude and shows that σ increases with rising degree of ionization (e.g., increasing temperature) and, since $\mu^- p$ is almost constant, decreases with increasing gas pressure. A hydrogen positive column at $p = 1$ mm Hg with $N = 10^{11}$ electrons per cm³ and $\mu^- = 4 \times 10^5$ cm² (V s)$^{-1}$ gives $\sigma \simeq 6 \times 10^{-3}$ (Ω cm)$^{-1}$ or a resistivity of 160 Ω cm.

When the temperature of the gas is sufficiently raised so that thermal ionization occurs, the number of collisions between electrons and ions becomes prominent. For a fully ionized plasma an approximate value of σ can be easily derived from equation 6.20. According to Langevin's theory of mobility of gases

$$\mu^- \simeq \frac{e}{m} \frac{\lambda}{\bar{v}} = \frac{e}{m} \frac{1}{(3kT/m)^{1/2}} \frac{1}{Nq} \tag{6.21}$$

where m and q are the mass of the electron and the collision cross-section for momentum transfer respectively and λ is the mean free path.

If an electron is strongly scattered by ions, i.e., through an angle $\pi/2$, when the ion potential energy $e^2/r_0 = 3kT$ (where r_0 is the distance of closest approach), then the scattering cross-section

$$q = r_0^2 \pi = \left(\frac{e^2}{3kT}\right)^2 \pi \tag{6.22}$$

and therefore from equations 6.20, 6.21 and 6.22 we find the conductivity of a fully ionized plasma

$$\sigma = \frac{(kT)^{3/2}}{e^2 m^{1/2}} \simeq 4 \times 10^{-5} T^{3/2} \qquad (6.23)$$

in $(\Omega \text{ cm})^{-1}$.

A logarithmic factor of value $\frac{1}{3}-\frac{1}{6}$ which results from the shielding effect is included in the numerical factor of equation 6.23. For a hydrogen plasma at $T = 10^4$ K this gives $\sigma = 40$ $(\Omega \text{ cm})^{-1}$. Note that σ is roughly independent of N because $\mu^- \propto 1/N$; N enters the shielding factor only logarithmically. A rise in plasma temperature increases σ; when T exceeds 10^6 K, σ becomes higher than that of metals at room temperature. Figure 6.10 shows how σ varies with N and T, in a fully ionized hydrogen plasma. The conductivity for 'electron circulation' is represented by σ^-.

This conductivity cannot be measured by inserting two electrodes in the plasma and passing a current through it unless the electrons which are removed from one electrode are replaced, for example by the other electrode. Otherwise positive and negative charges can only be extracted from the plasma at equal rates. Since in this case the (heavy) ions determine the rate of extraction, a plasma conductivity σ^+ is found (figure 6.10) which is at least 40 times smaller than that given by equation 6.23. The electronic conductivity σ^- has been measured by inducing currents in the plasma whereby electrons move on circular paths through their ion atmosphere and the ionic conductivity σ^+ by a cold probe method.

When a magnetic field acts on a completely ionized infinite plasma the conductivity depends on direction and is a tensor. If the current traverses the plasma normal to the direction of the field, the conductivity is found to be reduced

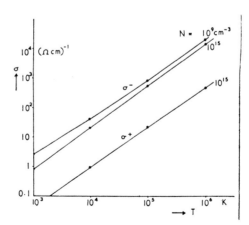

Figure 6.10. Electric conductivity σ of a fully ionized plasma as a function of temperature T and charge concentration N, in hydrogen.
The curves labelled σ^- represent the electronic conductivity for different N, and that labelled α^+ the ionic conductivity for H$^+$.

by up to one-third of its zero field value. The effect of the field is to change the energy distribution of electrons that are contributing to the current by reducing their average energy. This increases the number of slow electrons, the momentum transfer cross-section rises and the increase in effective collisions reduces σ, measured in the direction normal to that of the magnetic field. However, σ in the direction parallel to the field remains unchanged.

6.7. Thermal conductivity

It is known that the thermal conductivity κ of a gas in a normal state is given by

$$\kappa = (\bar{v}\lambda/3)\rho c_v \tag{6.24}$$

where the first factor is the diffusion coefficient D, and ρ and c_v are the gas density and specific heat at constant volume respectively. When a gas is feebly ionized, the contribution to heat conduction by electrons is very small. Since $D\rho$ is roughly proportional to $T^{1/2}$ and c_v increases with T, κ rises with T. However, in molecular gases at $T > 10^3$ K dissociation sets in and because of the energy required for it, $c_v = f(T)$ has a maximum beyond which the gas is atomic. Further peaks of κ occur when substantial amounts of energy are necessary for excitation (provided the process is confined to a narrow temperature range), ionization, etc. (figure 6.11).

The heat conductivity of a completely ionized plasma is found to rise strongly with T. As expected, κ is related to σ, the electric conductivity, by the Wiedemann–Franz law

$$\kappa/\sigma = C_1 T \tag{6.25}$$

and hence with equation 6.23

$$\kappa = C_2 T^{5/2} \tag{6.26}$$

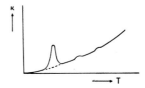

Figure 6.11. General variation of the thermal conductivity κ of a molecular gas as a function of the temperature T.
The first peak is due to dissociation, which happens at about 3500 K for H_2 and about 7500 K for N_2. The subsequent peaks are caused by excitation, ionization and other phenomena.

For metals, the constant C_1 is about $3k^2/e^2 = 2\cdot4 \times 10^{-8}$ V^2 K^{-1}, whereas for the plasma $C_2 = 4\cdot4 \times 10^{-9}$ V^2 K^{-1}. There are but few measurements.

We shall now compare the mechanism of unidirectional heat flow by conduction in a solid with that in an ideal plasma. Consider a metal rod between a hot infinite heat source at temperature T_1 and a cool infinite sink at T_2, both being electrically insulated. The heat flow from T_1 to T_2 proceeds by lattice vibrations (phonons) and electrons (de Broglie waves). Since the total electric current in the rod must be zero and electrons are forced to diffuse from T_1 to T_2, there will be an excess negative charge at T_2 and a corresponding electric field in the rod. Hence fast electrons of the distribution move down to T_2 against the field while slower electrons are accelerated by the field up to T_1; in the steady state the electric field is such that $i = 0$. Since the exchange in energy between electrons and ions in the lattice is feeble, heat is transferred mainly by the electrons.

In a fully ionized gas the conditions are very similar; again electronic heat conduction predominates. The picture given resolves an apparent paradox: the total specific heat of a solid is to a first approximation independent of that of the 'electron gas' because in the Fermi–Dirac distribution of energy there is only an insignificant number of electrons free to contribute to the specific heat c_v. It would be erroneous to conclude from equation 6.24 that, when $c_v \to 0$, κ is negligible since the electron density $N \propto \rho$ in metals is very high. As in the case of σ (section 6.6), κ is nearly independent of N because in equation 6.24 $\lambda\rho$ is approximately constant.

The thermal conductivity κ in a magnetic field B normal to the direction of heat flow is reduced by a factor of $1/(1 + \omega^2\tau^2)$ where $\omega = eB/mc \simeq 1\cdot8 \times 10^7 B$ (cyclotron frequency), B is in G and τ is the effective collision time.

6.8. Magnetic susceptibility

Consider a cylindrical metallic conductor which carries a current in an axial magnetic field B. Electrons are drifting slowly along the direction of B while moving simultaneously in circular paths around the field lines. The circular motion results from those electrons with a velocity component v_n normal to B, and occurs in a sense so as to reduce B. The circulating current would produce in the volume a diamagnetic moment were it not for the reflection of electrons from the outer boundary of the conductor which gives rise to a circulation of charge in the opposite sense. The corresponding paramagnetic moment turns out to cancel exactly the diamagnetic moment (figure 6.12(a)). This result is not unexpected: in a magnetic field the total energy of this system must remain the same if the electric field is to be zero.

If, however, a magnetic field acts on the plasma of a positive column at low pressure with both electrons and ions diffusing to the wall of the tube at the same time rate, then those electrons which reach the wall are neutralized by ions and no

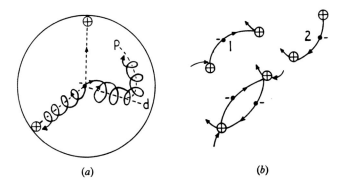

Figure 6.12. (a) Circular motion of electrons in a positive column in a field B normal to the paper, producing diamagnetic d and paramagnetic p effects, or diamagnetic effects only when wall recombination with positive ions takes place.

(b) Compounding electron paths in a field B (normal to the paper) interrupted by collisions with atoms (circles with crosses) which give rise to diamagnetic moments. The two 'half-loops' 1 and 2 are equivalent to a closed loop as shown below.

paramagnetic component of circulation can develop; the plasma bounded by walls exhibits diamagnetic susceptibility.

Assume that a uniform, axial magnetic field B is applied to a cylindrical, radially non-uniform, quasi-neutral plasma, such as a positive column of radius R, charge number density N, temperatures T_e, T_i and radial velocities v_r of electrons and ions moving towards the wall. Both charges are deflected by B, giving rise to tangential (angular) velocities v_{et}, v_{it} and a total angular current density $j_t = eN(v_{et} + v_{it}) = f(r)$. This is equivalent to a current encircling a cross-sectional area of the column which produces a magnetic field opposed to B and thus a diamagnetic moment $-M_d$. By superimposing on the d.c. current in the column a modulation current of about 1 kHz and using a cylindrical search coil surrounding the discharge tube to measure the coil voltage, the magnetic moment M_d per unit length of plasma can be derived.

The magnetic moment at low B originates from current elements of circular segments which, for reasons of symmetry, can be combined in pairs, thereby enclosing an area (perpendicular to B) with a current circulating along its circumference. The end of each segment is determined by a point at which an electron collides with an atom, which occurs v_e times per second, or every τ_e s. At high B, however, the charges move along circles, provided $v_e \propto p$ is low enough. It follows that now the areas enclosed grow first with B but decrease at higher B because the radius of gyration $\rho \propto 1/B$.

M_d per unit length of plasma (in A m) is, for N' charges per m length

$$-M_d \simeq \int_0^R eN_e \pi r^2 v_{et} \, dr = \frac{N'kT_e}{B} f\left(\frac{B}{p}\right) \tag{6.27}$$

where

$$f\left(\frac{B}{p}\right) = \frac{(\omega_e \tau_e)(\omega_i \tau_i)}{1 + (\omega_e \tau_e)(\omega_i \tau_i)} \propto \frac{(B/p)^2}{1 + (B/p)^2}$$

The shape of $f(B/p)$ is akin to a saturation curve. Inspection of equation 6.27 reveals that $M_d \propto B$ for low B, $M_d \propto 1/B$ for large B. Thus M_d has a maximum which can be found by differentiating it with respect to B. Note that $M_d \propto i$ through N'. Measurements in Hg confirm the theory quantitatively.

6.9. Acoustic effects

Any gas discharge in air can generate audible sound if an electric current of frequency one half of the sound frequency is superimposed on the d.c. discharge current. The reason that this method has so far only been used in special cases is the background electric noise, which has its origin in erratic fluctuations in the electrode zones as well as in the plasma region.

The 'speaking arc' was one of the earlier lecture experiments: a microphone circuit was inductively coupled to a d.c. carbon arc circuit which made the arc column emit ordinary speech; again, the unavoidable hum was troublesome. In principle, thermal adiabatic expansion and contraction in the gas produce longitudinal pressure and rarefaction, forming waves which travel from the acoustic source into the surrounding gas. If the discharge is at a temperature considerably above that of the ambient gas, then we expect the sound waves to travel through a hot, convectively moving gas whose temperature decreases with the distance from the source.

Instead of superimposing the sound current on the arc, the sound voltage can be applied to a pair of electrodes immersed in the plasma or to an induction coil co-axial with the (flame or discharge) plasma. Another powerful h.f. sound generator is the ionophone, which consists of a small cylindrical quartz cell with a conical opening. When a voltage of several kV at 30 kHz is applied between an external ring and an inner point electrode, an intense glow discharge in air is produced which can be modulated and the acoustic output can be either coupled to the ambient air or to water. In the latter case two ionophones, acting as transmitter and receiver of ultrasound waves in water, have been used to send television 270-line pictures of submerged objects across 50 m of water at a rate of 30 s^{-1}; see Klein (1974).

production of discharges and plasmas; diagnostics

7.1. General comments

The large variety of electric discharges makes it necessary to restrict the following account to a few typical cases which can be understood from first principles, using only data from the kinetic theory of gases, from atomic and molecular physics, and from thermodynamics. As to quasi-neutral plasmas, we shall distinguish between active ones, which require electric fields in order to be maintained, and passive plasmas, where charges of both signs are produced by irradiation, chemically or externally, their mixture being afterwards accomplished either *in vacuo* or in a gas-filled space; quasi-neutrality means that the net space charge is zero within a sufficiently large volume. The longitudinal electric field in the column of a partially ionized passive plasma is thus relatively low and so is the electron mean energy. It follows that the resulting electron distribution will not contain enough fast electrons to produce ionization by collision to balance the lost charges, and therefore they must be replaced as indicated above.

7.2. Evolution of a discharge

Consider the birth of a discharge in a long cylindrical glass tube which has plane cold metal electrodes at its ends and is filled with a gas at low pressure of order 1 Torr. The discharge is started by applying a variable voltage between the electrodes. A resistor in series with the d.c. source and the tube limits the current rising above an admissible value.

There are always some free electrons present, due to ionization by cosmic rays, radioactive impurities or background radiation. A charge multiplication process (sometimes called 'avalanche') in the field is always accompanied by electron scattering which deflects a small fraction of the (light) electrons towards the tube wall where they are held by electrostatic image forces or by surface atoms in the lattice, thereby forming (heavy) negative ions. This negative wall charge, which grows with time, is not neutralized by the heavy positive ions which arrive at a later stage. Soon, however, electron scattering towards the wall is reduced by the increasing negative wall charge, a process followed by acceleration of positive ions to the wall where neutralization between pairs of ions and electrons takes place. This process is similar to recombination of charges in the gas; but certain details of the motion of charges near and along the wall preceding neutralization are not yet fully understood. The neutralization energy gained is transferred to the electrons in the lattice and thus dispersed in the wall, while the newly formed neutral particles return to the gas. When a steady state is reached, an equal number of ions and electrons arrive per unit time on the wall while the bulk of the charges, which constitute the discharge current, move to the respective electrodes. Since charges in the gas are produced in pairs, the small electron wall charge is balanced by an equally small ion space charge which is distributed throughout the gas volume. This 'excess' charge gives rise to the radial electric field of a fully developed discharge and controls its radial loss of charge and of energy. However, the type of discharge that is finally established depends, as we

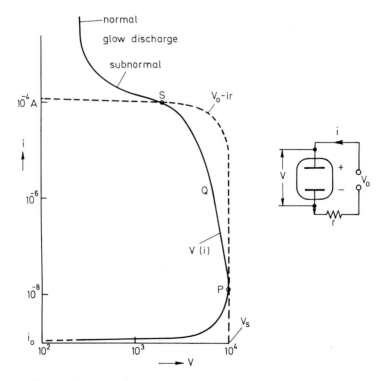

Figure 7.1. Change of current i from its small initial value i_0 in a gas at low pressure to that required to maintain a glow discharge where the source voltage V_0 is raised to the starting value V_s of the discharge.
At larger i the discharge voltage first decreases as i rises.

shall see later, on the constants of the circuit, the electrodes and gas parameters, and on the applied voltage.

How does the discharge current i vary when the voltage V applied between the electrodes is gradually increased? Figure 7.1 represents the answer. Let us assume that the cathode is irradiated with ultraviolet light so that it emits a small number of photo-electrons corresponding to an initial current $i_0 \sim 10^{-12}$–10^{-10} A. It is seen that i rises with V, first slowly and later faster, up to the point P, where $V = V_s$, the starting (or breakdown) voltage. Beyond P, however, V decreases while i rises. Obviously, P divides the 'characteristic' curve into two branches: for $V < V_s$ we find that $di/dV > 0$ and we have a non-self-sustaining discharge, since it can be stopped by switching off the ultraviolet light. Beyond P, $di/dV < 0$ and the discharge is self-sustaining. Irradiation is no longer needed and the discharge is self-perpetuating. Such a discharge starts when $V \geq V_s$.

It is easy to understand why i rises as V is increased: the ionization rate rises as the field $E = -dV/dx = f(V/d)$ in the gas is raised. The decrease in V with rising i is not so obvious. In this region, for cold electrodes, i becomes so large ($> 10^{-6}$ A) that E is no longer solely determined by the applied voltage V but by the net space charge. At low pressure, charges of both signs are lost to the vessel walls at a rate controlled by diffusion and the radial electric field. As a result of this, electrons and ions travel radially from the gas to the wall with a common radial speed, nearer to ion speeds, if strongly coupled, which is the case near Q in figure 7.1. Between P and Q, and further up, the charge loss to the wall gradually drops and hence a smaller E and V with a lower rate of ionization will keep the discharge steady. The current i can be found from the point of intersection S with the curve $V_0 - ir$, V_0 being the source voltage and r the external resistance, and the curve representing $V(i)$, the voltage required across the steady discharge. The steady discharges in figure 7.1 are termed, in order of ascending current, dark, glow and arc discharges, according to Faraday.

7.3. The Townsend discharge and the electric breakdown

Before turning to a detailed discussion of the 'Townsend discharge', which covers the region from 0 up to P, we note that in figure 7.1 V_s is of the order 10^4 V, since the example chosen refers to a medium gas pressure; at $p \sim 1$ Torr V_s may be much lower, say 100–500 V. Falling through the ionization potential V_i of 5–25 V gives an electron just enough energy to ionize a gas molecule, though the likelihood of ionization only becomes finite when $V > V_i$. The slow positive ions, unable to ionize molecules by collisions, are attracted by the cathode, from which they release secondary electrons which join and 'strengthen' the electron swarm which extends to the anode.

The condition for self-regeneration, as described before, is based on the following argument: when a single electron leaving the cathode is accelerated in the electric field, it produces m electrons and m positive ions in the gas on its way to the anode. When these m ions, plus the associated particles (metastables, quanta) release just one (secondary) electron at the cathode, the cycle considered is self-perpetuating and the ensuing discharge self-sustaining. Thus, if γ is the number of secondary electrons per ion striking the cathode the condition sought is simply

$$m\gamma = 1 \tag{7.1}$$

where m is the ion multiplication factor. Note that the first electron, unlike the rest, is not accompanied by a new ion. In the gas, the increase in number of electrons dn_e along dx in the field direction is proportional to the number of electrons n_e at x and a coefficient α, the electron ionization coefficient, in ion pairs per electron and unit distance. Hence at low i/i_0

$$dn_e = \alpha n_e \, dx \quad \text{and} \quad n_e/n_{e0} = i/i_0 = \exp(\alpha x) \tag{7.2}$$

The electron multiplication factor $\alpha = m + 1$, n_{e0} is the electron number at $x = 0$ and i_0 the 'initial' current. Hence from equation 7.1 and 7.2 with $x = d$, the interelectrode distance, $m = \exp(\alpha d) - 1$ and self-sustenance requires that

$$\gamma[\exp(\alpha d) - 1] = 1 \tag{7.3}$$

Hence, the bracketed expression describes the relevant gas properties and γ the secondary processes at the electrode–gas interface. From equation 7.3 the breakdown voltage V_s and field E_s can be numerically evaluated, provided the values of α (or α/p) and γ are known. From Townsend's semi-empirical relation (see figure 7.5)

$$\alpha/p = A \exp[-B/(E/p)] \tag{7.4}$$

p being the gas pressure (more precisely the gas density), and A and B constants. We remember that $E/p \propto E/\lambda$, the mean energy which an electron acquires when travelling along one electron mean free path in the field direction, which determines $\bar{\varepsilon}$, the mean electron energy of the swarm. The starting (or breakdown) voltage V_s for constant γ is

$$V_s = \frac{B(pd)}{\ln(pd) + \ln\{A/[\ln(1 + 1/\gamma)]\}} \tag{7.5}$$

showing that its value depends on pd only, the number of molecules in the plane-parallel gap, and not on p or d separately. Equation 7.5 represents the 'Paschen curve' (figure 7.2) which has a minimum at low pd (< 1 cm Torr) of about $V_{s\,min} = 300$ V. Thus, below that voltage, except for special cathodes with large γ and certain rare gases, a self-sustaining discharge cannot be started. For air and H_2, $A = 15$ and 5, $B = 360$ and 140 respectively in (cm Torr)$^{-1}$ and V (cm Torr)$^{-1}$.

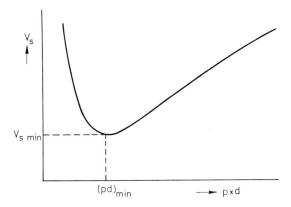

Figure 7.2. Starting voltage $V_s = f(pd)$ in the low p range (Paschen curve). p = equivalent gas density.

When two disc electrodes filling up the diameter are fixed a short distance (order 10 mm) apart in a glass torus (discharge vessel) of axial length ~ 500 mm and 5–10 mm inner diameter that contains gas at low p (≤ 0.1 Torr) a discharge in this 'detour tube' can be started by applying a voltage of <10 kV to the electrodes. However, a light emitting discharge can only develop in the long section of the torus but not in the short electrode gap where, owing to the low pd (figure 7.2), the required starting potential is too high to initiate a discharge.

The breakdown voltage in equation 7.5 rises above the minimum as pd is increased and the corresponding field $E_s = V_s/d$ varies approximately linearly with p, since the denominator of equation 7.5 changes slowly at high p. At 1 atm $E_s \sim 30$ kV cm^{-1} for air and ~ 20 kV cm^{-1} for H_2 for plane electrodes.

Returning to the Townsend discharge, equation 7.2 can be written as $\ln(i/i_0) = \alpha x$ and tested by measuring the current in the circuit. When V is very low no multiplication occurs in the gas and $i \sim i_0$, but when V is large so that the gas in ionized and $x = d$, equation 7.2 holds. The value of α can therefore be derived from the slope of the curve $\ln(i/i_0) = f(x)$ where E/p is kept constant as shown in figure 7.3; for a sufficiently large value of E/p the curve deviates from a straight line at larger x; analysis of this section gives the required value of the secondary electron coefficient γ. Another point of interest is that $\alpha/p = f(E/p)$ is closely related to the ionization cross-section (see Chapter 3) because α/p depends ultimately on $\bar{\varepsilon}$. Hence only electrons with energies $\varepsilon \geq eV_i$ (in the energy distribution) contribute to ionization via $q_i = f(\varepsilon)$ (see figure 3.7, p. 42). In He, Ne, Ar, H_2, N_2, O_2, Hg and air the minima for Fe are at about 220, 230, 300, 250, 230, 480, 450 and 420 V, and at 3, 5.5, 1.8, 1, 0.5, 0.4, 1 and 0.7 cm Torr, respectively.

The electron current $i = i_e$ is collected by the anode, whereas at the cathode, apart from a small secondary electron component, $i \simeq i^+$. The spatial distribution of the two current components is seen in figure 7.4, indicating that at any point x

Electric plasmas: their nature and uses

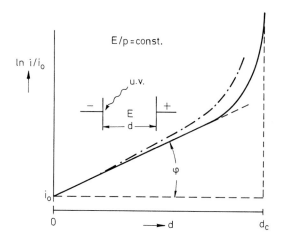

Figure 7.3. Current multiplication $i/i_0 = f(d)$.
The electrode separation for E/p (or E/N) = constant in a parallel plane gap;
——— = molecular; — · — · — = rare gas; $\tan \phi = \alpha$.

between 0 and d the total current i is constant, $i = i_e(x) + i^+(x)$, the charges n_e and n^+ being related to i_e and i^+ by the respective drift velocities (see Chapter 4). We mentioned earlier that the electrons move swarm-like and can often be described by their mean energy $\bar{\varepsilon}$. This quantity too has been measured by subdividing the anode into concentric rings and observing the electron current entering each individual ring. This gives $i(r)$, the radial distribution of charge, and since the radial spread is due to radial diffusion of electrons in the gas, which in turn depends on the electron temperature T_e or the mean energy $\bar{\varepsilon} = kT_e$, $\bar{\varepsilon}$ can be derived from $i(r)$, if the electron drift velocity $v_d = f(E/p)$ is known.

The role of the positive ions here is simply that of slow-moving charge carriers. At the cathode, the total current was said to be virtually a positive ion

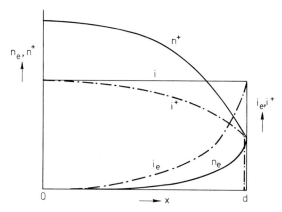

Figure 7.4. Spatial distribution of the current components i_e and i^+, and of the charge concentrations n_e and n_i^+.

current. This suggests that positive space–charge effects will develop first at the cathode if the currents become sufficiently large, which indeed is the case.

The validity range of Paschen's relation

Carefully conducted experiments revealed that equation 7.5 applied near and not too far above the minimum, but failed at lower and higher *pd* values. Townsend assumed that V_s depends on the number of molecules Nd in a gap *d* per unit area for a uniform field of relative value E/N and constant ionization coefficient α/N. At high *p* or *N* strong space charges in the multiplying electron swarms will distort the local field *E*, a fact that was recognized some 40 years ago when it was shown theoretically that the formative time lag strongly decreases when *p* is raised. However, though this agreed with the oscillographic results the theory did not include the final stage of the discharge—the spark—which is closely related to a temporary arc. Today's physical picture differs in several respects. It operates with a non-uniformly distributed space charge which develops in the head section of the electron 'avalanche'. There, at moderately strong ion concentrations, the local field is reduced at the anode side and raised at the cathode side of the head from which the electrons are rapidly removed, while the ions remain virtually stationary. At high ion concentration a plasma filament is formed because the numerous slow electrons are now kept back by the ions except for faster electrons which occupy a thin sheath near or at the anode. As a result of this a low or zero field plasma region appears. The former voltage drop along it serves to activate the avalanche section on the cathode side which it converts into a plasma canal. In this way a cathode-directed plasma stem grows (see Further Reading, Chapters 7 and 12, lightning). Note that the last model does not assume that breakdown relies on ionizing u.v. radiation in the gas. We conclude that equation 7.5 fails at low and high gas densities where V_s depends on *d* but not on *pd*; at higher *p* electron field emission becomes important. V_s then depends on the cathode material and its surface layer (see von Hippel, 1965).

7.4. Ionization coefficient and ionization cross-section

The electron ionization coefficient α is of importance in all cases where an electric field *E* and thus a mean electron energy $\bar{\varepsilon}$ for a specific energy distribution is given, so that the rate of ionization, per colliding electron per second, is

$$z = \alpha v_d \qquad (7.6)$$

v_d, being the electron drift velocity, can be determined. Since α is proportional to the number density N_0 of molecules at constant *T*, it is convenient to write $z/p = (\alpha/p)v_d$; observations of $\alpha/p = f(E/p)$ are shown for a few gases in figure

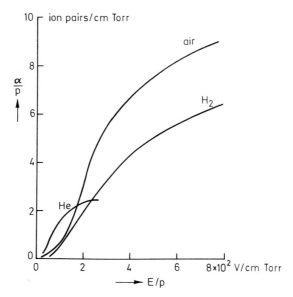

Figure 7.5. Reduced ionization coefficient α/p as a function of the reduced field E/p for various gases at 300 K.

7.5. Treating ionization as a chemical reaction the reaction rate, in ion pairs ($= 1$ electron $+ 1$ positive ion) per second per unit volume, is

$$zN_e = dN_e/dt = CN_e N_0 = \overline{q_i v_e} N_e N_0 \tag{7.7}$$

where C, the rate constant, equals the average of the product ionization cross-section $q_i = f(\varepsilon)$ and electron velocity v_e ($\simeq$random velocity). This average $\overline{q_i v_e}$ is found by integrating over all velocities of the known energy distribution $f(\varepsilon)\,d\varepsilon$. Hence, with the electron concentration N_e and the molecule concentration N_0

$$\frac{dN_e}{dt} = N_e N_0 \int_0^\infty q_i(\varepsilon)v_e(\varepsilon)f(\varepsilon)\,d\varepsilon \tag{7.8}$$

(To write instead for the lower limit $\varepsilon = \varepsilon_i$ is irrelevant, since $q_i = 0$ between $\varepsilon = 0$ and $\varepsilon = \varepsilon_i$.) For $q_i v_e$ we can write, with q_{tot}, the total collision cross-section $= 1/\lambda_e N_0$

$$q_i \lambda_e v_e/\lambda_e = \left(\frac{1}{N_0}\right)\left(\frac{q_i}{q_{tot}}\right)\left(\frac{v_e}{\lambda_e}\right)$$

Note that the three factors under the integral in equation 7.8 are: q_i/q_{tot}, the reduced ionization cross-section ($\propto$likelihood of ionization); v_e/λ_e, the electron collision frequency; and $f(\varepsilon)\,d\varepsilon$, the fractional number of electrons in $d\varepsilon$, often represented by the Maxwellian distribution (see figure 3.2(*a*), p. 33). It can be readily shown that since all N_e electrons must be found to have $0 \leq \varepsilon \leq \infty$, then

$$\int_0^\infty f(\varepsilon)\,\mathrm{d}\varepsilon = 1$$

Thus equation 7.8 states that the ionization rate per unit volume and per second is

$$\frac{\mathrm{d}N_e}{\mathrm{d}t} = N_e \int_0^\infty (q_i/q_{\text{tot}})(v_e/\lambda_e)f(\varepsilon)\,\mathrm{d}\varepsilon \tag{7.9}$$

The type of distribution can be Maxwellian, i.e., with $\bar{\varepsilon} = kT_e$

$$f(\varepsilon)\,\mathrm{d}\varepsilon = \frac{2N_e}{(\pi)^{1/2}} \left(\frac{\varepsilon}{\bar{\varepsilon}}\right)^{1/2} \exp\left(-\varepsilon/\bar{\varepsilon}\right) \tag{7.10}$$

Yet it is often found that the tail of $f(\varepsilon)$, i.e., the region where $\varepsilon > eV_i$ (figure 7.6), contains fewer electrons than is given by equation 7.10. The energy distribution which applies in a specific case depends on the reduced field E/p (or E/N_0) and on the nature of the gas, in particular on its Ramsauer–Townsend effect $q_{\text{tot}} = f(\varepsilon)$ (see figure 3.6, p. 40), and at larger j, on stage processes, but this is beyond our scope. From equations 7.6, 7.7 and 7.9 we obtain the relation between α and q_i, namely

$$\frac{\alpha}{N_0} = \frac{1}{v_d} \int_0^\infty q_i (2\varepsilon/m)^{1/2} f(\varepsilon)\,\mathrm{d}\varepsilon \tag{7.11}$$

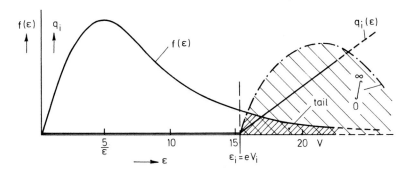

Figure 7.6. Illustrating the ionization rate integral, equation 7.8.

With $\lambda_e = \lambda_1/p$ and $v_e/\lambda_e \simeq \overline{pv_e/\lambda_1}$ for constant T_{gas}, one can approximate equation 7.11

$$\frac{\alpha}{p} \sim \left(\frac{1}{v_d}\right)\overline{v_e/\lambda_1} \int_0^\infty (q_i/q_{tot})f(\varepsilon)\,d\varepsilon \qquad (7.12)$$

with v_d depicted in figure 4.24 (p. 77), λ_e in figure 3.6 (p. 40), q_i in figure 3.7 (p. 42) and $\bar\varepsilon = f(E/p)$ in figure 7.7. The integral equation 7.12, drawn schematically in figure 7.6, the area under the dash-dotted curve, is the number of electrons in the tail of the distribution represented by the small shaded area between $\varepsilon_i = eV_i$ and $\varepsilon_i = \infty$; the ionization cross-section $q_i(\varepsilon)$ (the part near the onset) can usually be approximated by a straight line. Equation 7.12 has been found to hold in all cases in which the distribution is known and stage processes are absent.

7.5. Genesis and properties of a glow discharge

This discharge derives its name from a luminous region near the cathode, the visual extent of which depends on the gas pressure (the larger p the shorter its axial length) and the current density. Figure 7.8 shows the dark and luminous 'spaces', assuming that cold plane metal electrodes are used in a cylindrical vessel of a few cm radius filled with a gas at a pressure of $0 \cdot 1$–1 Torr and a current of the order 1 mA. In pure rare gases simpler conditions are encountered than in molecular gases, where band spectra appear and dissociation complicates the

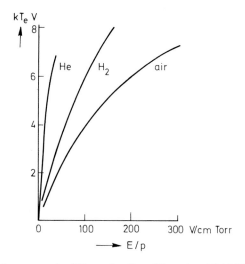

Figure 7.7. Mean electron energy $\bar\varepsilon = kT_e$ as a function of the reduced field E/p for various gases.

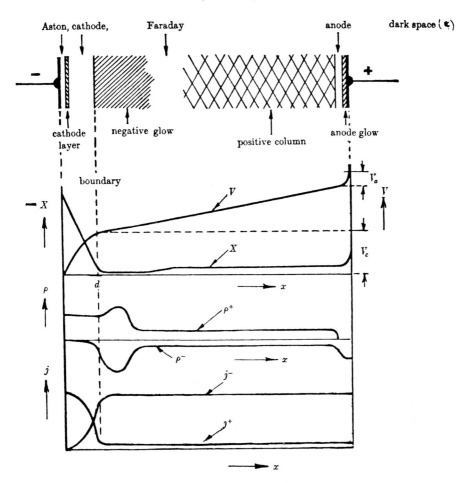

Figure 7.8. V_x, E_x, ρ^+, ρ_e, j^+, $j_e = f(x)$ in a cold cathode at low pressure and with normal flow discharge.

situation. Apart from structural details, of main interest are the cathode and Faraday dark spaces, separated by the negative glow. The positive column fills the remaining space between Faraday's dark space and the anode, ending in the anode glow. This either covers the anode in the form of a layer or it consists of 'individual' luminous spheres 'stuck' to the anode surface; their origin is still not clearly understood.

If a gas discharge is started by slowly increasing the external voltage (instead of pre-ionizing the gas by a Tesla coil which induces a high frequency discharge), the first stage in the development is the dark discharge (figure 7.9), the stabilization of which requires very large applied voltages. The voltage across the dark discharge V_d decreases in general with the current density according to

$$V_d = V_s - \text{constant } j^2 \tag{7.13}$$

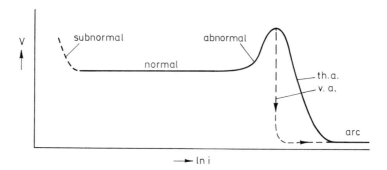

Figure 7.9. Three different modes of glow discharge dependent on the current i.
th.a. = thermal arc; v.a. = vapour arc.

since a field distortion ΔE by a space charge ρ would reduce V_s, the starting potential by an amount $\Delta V \propto j$ if $d^2\alpha/dE^2$, the curvature of $\alpha = f(E)$, were zero; but a positive curvature makes $\Delta V \propto j^2$. (The proof will not be given here.)

The normal glow discharge

When the current i in a discharge of moderate length is raised further, the discharge voltage first decreases rapidly and then at a slower rate until it reaches a constant voltage, the 'normal' cathode fall V_n (figure 7.9). The (dashed) transition range is termed the 'subnormal' regime. At the low current side of V_n, the region of strongest luminosity is the negative glow which covers only part of the cathode surface. V_n remains constant even when i is increased by over two orders of magnitude. (This property of the normal glow discharge has been used to produce constant voltage sources.) However, with rising i the cathode area covered by the glow expands proportionally, i.e., the cathode current density j_n remains constant. By varying the dimensions of discharge tubes it was found that j/p^2 is a similarity parameter of the cathode, meaning that cathodes of the same material whose parameter has the same value (whatever i and p) produce similar discharges, i.e., have the same normal and abnormal cathode falls V_n and V_c, as will be seen.

The constancy of V_n for large variations of i requires a physical explanation of the 'force' that keeps the almost dark cylindrical negative 'column' stretching between the negative glow and the cathode from expanding or contracting radially. The 'radial cohesion' is thought to be due to E_r, the radial component of the electric field of the ion-filled column. Figure 7.10 gives the parabolic potential distribution $V_{r=0} = f(x)$ along the dark column's axis, found by integrating equation 7.14, and $V = f(x)$ in the charge-free space outside the column, which is linear. It is seen that, except for $x = 0$ and $x = d_n$, there is a finite potential difference between $V_{r=0}$ and V at any x, causing a radial field which accelerates the positive ions outwards. This effect is equivalent to ion self-repulsion. If i and V are given, a 'virtual' rise in column radius R_c would reduce E_r and thus cause

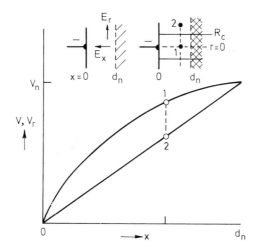

Figure 7.10. Local potential V along x, the axis d in the core and along x in 2, outside the core of the discharge (R_c) in the cathode dark space of a glow discharge of cathode fall V_n. (A. von Engel, K. G. Emeleus and M. Kennedy, 1972, *Physics Letters* **42A**, 191.)

the ion beam to contract, whereas a virtual reduction in R_c would widen the beam spot at the cathode. This argument demonstrates the stability of the negative column radius against a virtual radial change, and the constancy of the cathode current density; it expresses the simultaneous actions of the longitudinal and radial fields, including the ion scattering in the neutral gas. Electrons do not affect the fields, except near the dark space-glow boundary, because of their low number densities.

In table 7.1 values are given for various combinations of the gases and cathode substances which include the thickness d_n of the cathode dark space which varies with $1/p$ at constant T_{gas}. Thus pd_n is another similarity parameter, which for a long time was used to estimate the gas pressure in vacuum systems filled with a known gas.

By observing the deflection of an electron beam shot across the dark space, or from the Stark effect of atomic lines, the electric field E in the x direction is

Table 7.1. Values of the normal cathode fall (V_n), the cathode current density (j_n/ρ^2) and the dark space (pd_n) in normal glow discharges. Values for air and N_2 are similar.

Gas	Cathode	$V_n(V)$	$\dfrac{j_n/\rho^2}{(10^{-6}\,\text{A cm}^{-2}\,\text{Torr}^2)}$	$\dfrac{pd_n}{(\text{Torr cm})}$
He	Ni or Pt or Fe	~160	2	1·5
Ne	,,	~160	6	0·9
Ar	,,	130	160	0·3
H_2	,,	250	70	0·9
N_2	,,	210	400	0·4
Hg	Hg or Fe	300	8	0·3

found to decrease linearly with x (figure 7.8). This is consistent with a net space charge $\rho_{net} = \rho^+ = $ constant, which means that electrons do not contribute to the space charge in the dark space, except near the negative glow. Hence from Poisson's equation (neglecting the radial component E_r)

$$\frac{dE}{dx} \sim \frac{\rho^+}{\varepsilon_0} = \frac{j^+}{\varepsilon_0 v_d^+}; \quad E/E_0 = 1 - x/d_n \qquad (7.14)$$

where d_n is the distance between the cathode ($x=0$) and the 'boundary' d_n, and E_0 the field at $x=0$. At $p=1$ Torr, in a molecular gas, $E_0 \sim 10^3$ V cm^{-1}, $\rho^+ \sim 10^8$ ion cm^{-3}. Hence, $V_n = \int_0^{d_n} E \, dx$, which is the area of the triangle given by equation 7.14. Thus, $V_n = \frac{1}{2}E_0 d_n$ and with $d_n \sim 0.5$ cm, $V_n \sim 300$ V.

The normal cathode fall V_n (table 7.1) is shown to be dependent on the gas and the cathode material. It can, however, be related to the corresponding collision parameters. If the cathode and its dark space constitute a self-sustaining system then, instead of $\gamma[\exp(\alpha d) - 1] = 1$ as for constant E, E and α vary here with x (equation 7.4 and 7.14) and

$$\int_0^d \alpha \, dx = \ln(1 + 1/\gamma) \qquad (7.15)$$

provided α/p is still meaningful. Replacing the integral by $\alpha d = \eta V_n$ where η is the average number of ion pairs per V produced by an electron in the gas, i.e., $\eta = \alpha/E = \overline{\alpha}d/V = (\alpha/p)/(E/p)$, we can rewrite equation 7.2

$$i/i_0 = \exp(\eta V) \qquad (7.16)$$

with $(1/\eta)$ of the order 50 V per ionization, and obtain from equation 7.15

$$V_n = \overline{\alpha}d/\eta = (1/\eta)\ln(1 + 1/\gamma) \qquad (7.17)$$

Hence V_n is the lower the larger η or α, or q_i or V_i, and the larger γ. It is not surprising, therefore, that with a caesium-coated nickel cathode in neon a value as low as $V_n \sim 40$ V has been observed, though with pure metal cathodes $V_n \sim 150$–250 V.

An estimate of (pd_n) follows from equation 7.15 and the relations $\bar{\alpha}d_n = (\bar{\alpha}/p)(pd_n) = \ln(1 + 1/\gamma)$

$$pd_n \simeq \frac{\ln(1/\gamma)}{(\bar{\alpha}/p)} \simeq \frac{\ln(1/\gamma)}{\eta V_n} \qquad (7.18)$$

Thus pd_n rises as γ and $\bar{\alpha}/p$ decrease (e.g., in helium). Finally, j_n/p^2 is found when we put $i \simeq i^+$ at the cathode. From equations 7.14, 7.15 and 7.18 one finds that the 'reduced' normal cathode current density

$$j_n/p^2 = \frac{2(\mu^+ p)(\bar{\alpha}/p)}{\eta} = \frac{\mu^+ \ln(1/\gamma)}{\eta d_n} \qquad (7.19)$$

In light rare gases and in mercury j_n/p is small because $\bar{\alpha}$ is small, η relatively large and μ^+ (actually the 'charge transfer mobility') small; in molecular gases j_n/p^2 is up to about 100 times larger, since $\bar{\alpha}$ is larger and η smaller than in atomic gases. Note that the right hand sides of equations 7.17, 7.18 and 7.19 are independent of p, the equivalent gas density. The values of γ are of the order 10^{-1}–10^{-3} with ordinary metals in atomic or simple molecular gases, but very low values, down to 10^{-10}, have been observed with large organic molecules whose ions have a large number of degrees of freedom, not describable by simple vibrations and rotations. Their impact with the cathode excites the neutralized molecules, thereby increasing its internal energy, rather than liberating a second electron from the cathode (the first neutralizes the ion). Equations 7.18 and 7.19 have been expressed in two ways so as to avoid misintepretation: obviously each of the three cathode parameters must contain a quantity related to gas ionization and another to the gas–solid interface emission, and inspection of the relations confirm this. However, since in such a self-sustained system the reproduction condition of equation 7.15 must hold, factors containing the coefficient γ describing the solid–gas interface can formally be replaced by parameters hiding γ.

7.6. The abnormal glow discharge

When the current is increased beyond the value at which the glow covers the whole cathode surface, then a larger current requires a cathode fall $V_c > V_n$. Figure 7.11 gives the abnormal cathode fall $V_c = f(j_c/p^2)$ for an iron cathode in various gases. At the same time the 'reduced' dark space value (pd_c) is constant as long as $V_c = V_n$, begins to decrease, i.e., the dark space shrinks (axially)

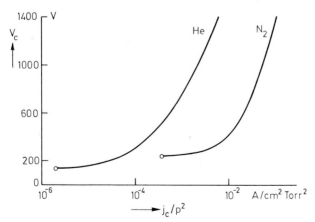

Figure 7.11. Transition from the normal $V_n(o)$ to the abnormal cathode fall V_c in He and N_2 on Fe as a function of the normalized cathode current density j/p^2.

whereby the region where most of the positive ions are produced moves from the dark zone near the boundary into the negative glow.

The characteristic $f(V_c, j_c/p^2)$ results from the following relations (in S.I.)

$$j_c = (j_e + j^+)_c = j_c^+ (1 + \gamma)$$

$$j_c^+ = (\rho^+ v_d^+)_c; \rho^+ = \varepsilon_0 \frac{E_0}{d_c} = \frac{2V_c}{d_c^2} \varepsilon_0$$

Hence, with v_d^+ from p. 74

$$j = \frac{2v_d^+ V_c \varepsilon_0}{d_c^2} (1 + \gamma) \tag{7.20}$$

if $\gamma \ll 1$ and $v_d^+ \propto (E_0/p)^{1/2}$. In similarity terms this reads, for $\mu^+ \sim$ constant

$$j_c/p^2 \propto V_c^{3/2}/(pd_c)^2 \tag{7.21}$$

and its value is independent of the gas density. $(pd_c)^2$ can be eliminated by a reproduction equation, similar to equation 7.3. Assume that ions are produced in the dark space d_c by a swarm-like electron component plus a fast component of energy of about V_c. If the fraction a of all electrons represents the swarm, then $(1 - a)$ are fast. The reproduction relation thus reads

$$a \exp\left(\frac{\bar{\alpha}}{p} pd_c\right) + (1 - a)\left\{\exp\left(\frac{\bar{\alpha}}{p} pd_c\right)\right\} V_c/z \sim 1/\gamma \tag{7.22}$$

so that $pd_c = f(V_c)$ follows from

$$pd_c \sim \frac{1}{\bar{\alpha}/p} \ln\left(\frac{1/\gamma}{a + (1 - a)V_c/z_g}\right) \tag{7.23}$$

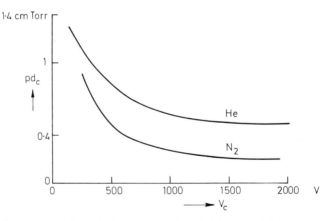

Figure 7.12. Contraction of the cathode dark space d_c with rising cathode fall V_c for He and N_2 on Fe.
For N_2: $V_c \sim 300$ V; $v_d^+ \sim 5 \times 10^3$ m s^{-1}; $d_c \sim 6$ mm at 1 Torr; $j \sim 1$ A m^{-2}.

where $\bar{\alpha}/p = f(\bar{E}/p)$; $a = 1$ for $V_c = V_n$ and decreases as V_c rises; and z_g is the total number of ion pairs per V in the glow (see Chapter 4). Figure 7.12 shows $pd_c = f(V_c)$ where p is the gas density in equivalent Torr. Inserting equation 7.23 in equation 7.20 gives

$$j_c/p^2 \propto V_c^{3/2} \left/ \left[\ln \left(\frac{1/\gamma}{a + (1-a)V_c/z_g} \right) \right]^2 (\bar{\alpha}/p)^2 \right. \tag{7.24}$$

From figure 7.11 it follows that a considerable heating of the gas in the cathode dark space occurs, so that its 'equivalent' pressure is lower than that read on a pressure gauge. Consider the dark space of normal glow discharges at $p = 1$ atm in air and H_2 where the size of the cathode spots at copper electrodes is easily measurable. $(j_c)_{1\,atm}$ is found to be about 8 and 5 A cm^{-2} respectively. Since $(j_c/p^2)_{air}$ is constant, 10^{-2} A cm^{-2} at 10 Torr, j_c at 1 atm in air would be expected to be 64 A cm^{-2} for constant T_{gas}. However, calculations have shown that, for example, at 1 atm in the dark space, for watercooled cathodes, $\bar{T}_{gas}$ is about 1000 K and about 700 K in air and in H_2 respectively, so that theory and observations agree when the change in gas density is allowed for.

Only two other forms of the glow discharge will be briefly discussed: the corona and the hollow cathode discharge. On wires of electrofilters (electrostatic precipitators) for removing small dust or smoke particles, in ozonizers and on power lines in open air a corona discharge develops when the electric field at the conductor surface exceeds a critical value. The atmospheric (negative) corona is a highly non-uniform glow discharge, the cathode end of which forms luminous spots of large current density, whereas the surrounding space bounded by the anode remains dark. If V_0 is the onset potential (see Chapter 4), the corona current is $i \propto cV(V - V_0)$ where c depends on the average ion mobility (see figure 4.23). Since p is large, the luminosity (observed with a microscope) is entirely confined to the negative glow, and the Faraday dark space (as in the atmospheric glow discharge) is distinguishable, but not the small cathode dark space. The chemical activity of the negative glow is significant (a.c. corona; Cobine, 1941).

The low-pressure hollow cathode discharge, a spectroscopic light source of high emission efficiency, low power consumption and small line width through Doppler broadening, has often a hollow cylindrical cathode whereas the shape and position of the anode is of little importance (figure 7.13(a)). The mechanism is best understood when a plane double cathode (figure 7.13(b)) replaces the cylinder. The simple facts emerge from figure 7.14 where the dashed curve shows observations of $i_1 = f(p)$ with a single cathode at $V_c = 400$ V in xenon, in agreement with theory (equation 7.21): since for constant V_c, j_1/p is constant, the current of the abnormal discharge, fully covering the cathodes, is $i_1 \propto p^2$. With both cathodes, distance D apart, the total current $i_2 = f(p)$ is obtained, which shows a maximum at a certain value of p, demonstrating the 'hollow cathode effect'. We note (figure 7.15) that $i_2/i_1 = f(pD)$ can rise up to several hundred, depending on the gas and V_c. The effect occurs in atomic as well as in molecular

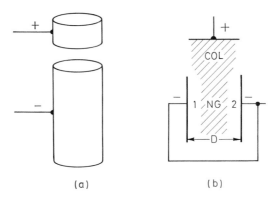

(a)   (b)

Figure 7.13. (*a*) Cylindrical hollow cathode HC.
(*b*) Plane hollow (double) cathode 1, 2 with negative glow NG and positive column COL (for details see Further Reading).

gases. The main cause of the large multiplication i_2/i_1 seems to be that in the ordinary glow discharge the fast electrons proceed through the glow to the anode, whereas here, after having traversed the glow, they enter the cathode dark space of the second and, after retardation, are accelerated again and 'repelled' into the negative zone of the first cathode. This pendulum motion interrupted by scattering ceases when the energy loss which the electrons suffer has reduced their velocity to a level at which the anode field is strong enough to extract them from the gas.

The anode fall in glow discharges

The anode acts on the plasma of the positive column by repelling the positive ions. This sets up a thin sheath of electron space charge from which is drawn the

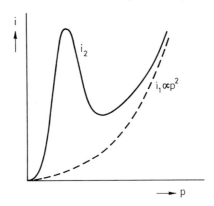

Figure 7.14. Total current $i_1 = f(p)$ at constant V_c for large D (HC effect absent) and $i_2 = f(p)$ for low p and D when a large value of i_2 is obtainable.

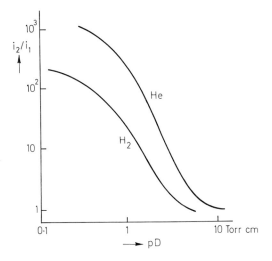

Figure 7.15. Current multiplication of i_2/i_1 in an He and H_2 plane hollow cathode system dependent on pD.

electron current. This fall region, causing the anode fall V_a, usually of order V_i, has been verified by electron beam deflection measurements. At larger i such as in arc columns, $V_a < V_i$, and at 'point' anodes V_a can rise to several hundred at small i. At lower p the anode is uniformly covered by an anode glow, which often contracts, thereby forming an array of symmetrically distributed luminous spheres (see Further Reading).

7.7. Charge and energy losses in the positive column at low pressure

Number loss of charge and electron temperature

A uniform electric plasma can be part of an electric discharge which extends between two electrodes, it can be maintained in an electrodeless gas discharge, or it is caused by chemical ionization such as in a gas flame. Its constituent charges are lost either after contact with the electrodes, whereby the 'lost' charge circulates through the attached circuit, or by recombination of charges when opposite pairs combine in the volume of the gas or neutralize each other on the walls of the vessel. In all these cases neutral particles are formed. Dust particles act often as 'microwalls' if their size is large, otherwise 'heavy ions' are produced.

Obviously, the life of an electron or ion, that is, the average time counted from the moment a free charge is born until it disperses or is immobilized, is an important factor. Therefore, to keep the electron population in a steady state, the average life of a free electron (in s) times the average rate of ionization, i.e., the number of ion pairs created per second, must be unity, provided that ionization occurs directly, i.e., by single electron collisions. Of course, the same condition must apply to other processes and groups of particles. Every new electron created in a gas is accompanied by a new positive ion, and hence in the steady state the life and rate of production of ions must equal that of electrons, provided the gas is the only source of ionization. However, this does not apply when, for example, positive ions are released from a surface by, say, thermionic emission or strong electric fields, because ions and electrons are, in general, not emitted in equal numbers. Or, if negative ions are formed by electrons that become attached to neutrals, the mean life of the negative charges is increased above that of electrons, and their loss (and ionization) rate is diminished.

Consider a cylindrical column of a partially ionized gas enclosed in a tube containing electrons, positive ions and neutral molecules. Let the rate of ionization be determined by single collisions between electrons and molecules with concentrations N_e and N_0 respectively. The ionization rate can be expressed by dN_e/dt (in ion pairs per cm^3 per second) or z (in ion pairs per second per electron)

$$\frac{dN_e}{dt} = \frac{dN_i}{dt} = zN_e = \overline{q_e v_e} N_e N_0 \tag{7.25}$$

where the rate constant $\overline{q_e v_e} = f(\bar{\varepsilon})$ depends on kT_e and the gas used. The average life τ_e of an electron is given by the time it takes it to diffuse, say, from the axis to the wall where it neutralizes a positive ion without delay. For ambipolar diffusion (see figure 4.26, p. 82) the coefficient $D_a = f(T_e)$ and the tube radius R determine τ_e, given by the random walk equation for one dimension $x^2 = 2Dt$. Thus

$$\tau_e \simeq \frac{R^2}{2D_a} \tag{7.26}$$

In the steady state we have therefore (see above)

$$z\tau_e = 1 \tag{7.27}$$

and from equations 7.25 and 7.26

$$\overline{q_e v_e} N_0 \simeq 2D_a/R^2 \tag{7.28}$$

where $D_a N_0$ is independent of p at constant T. Since $D_a \sim \mu^+ kT_e/e$ and $\overline{q_e v_e} \propto \exp(-eV_i/kT_e)$, T_e can be obtained from equation 7.27. Figure 7.16 shows the general curve $T_e/V_i = f(cpR)$ where $c = (aV_i^{1/2}/\mu_i^+)^{1/2}$; a is the slope of $Q_{i1} = q_i N_{e1}$ in cm^{-1} at $p = 1$, V_i is in V, μ_i^+ in cm^2 V^{-1} s^{-1} at $p = 1$. For Ne, $c = 5 \cdot 9 \times 10^{-3}$. At low and high pR, the general curve does not apply any longer,

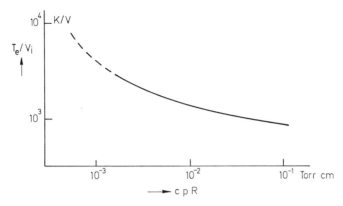

Figure 7.16. Electron temperature T_e (reduced to unit ionization potential V_i) dependent on gas c, pressure p and column radius R.

since at low p the ambipolar concept fails so that as pR decreases T_e rises less steeply than shown. At larger p, electron–ion recombination sets in and equation 7.26 has to be replaced by a condition in which

$$1/\tau \simeq 2D_a/R_c^2 + \rho_r N_e \tag{7.29}$$

where $R_c < R$ is the column radius, R that of the vessel, ρ_r the recombination coefficient in cm^3 s^{-1}. The mechanism of contraction of positive columns is not yet fully understood. Also at large p and i similarity relations involving pR, E/p, j/p^2 and so on, fail. Expressions of the form of equation 7.27 apply to steady states as well as to breakdown although, for example, the breakdown potential, V_s, is always higher than that for maintenance, V_m. This apparent contradiction is resolved thus: in the case of V_s the electrons are lost by free electron diffusion D_e, whereas the steady-state loss is controlled by D_a, i.e., ambipolar diffusion, and sometimes by recombination. Note that $D_e \gg D_a$. At lower i, the positive column parameters T_e and E/p are independent of i or j. At larger i, electrons ionizing excited atoms and molecules, collisions between excited particles (associative ionization) giving molecular ions, and photo-ionization have to be included. Thus T_e and E/p become dependent on i. Single collision ionization is then replaced by ionization in stages. At larger i changes in gas temperature and density must be considered.

The losses in a positive column maintained at very high frequencies (h.f.) are not substantially different from a d.c. column, because $\bar{\varepsilon}$ is hardly modulated at larger p though E varies with time. This is so as long as the electron motion is frictional, i.e., when the inverse mobility term 4.35 is large compared with the inertial or mass term: $e/\mu_e \gg m\omega^2$. In this case the electron velocity and the current are in phase with the field $E_t = E_0 \sin \omega t$. This is equivalent to the situation when the applied frequency $\omega < \nu$, ν being the electron–molecule collision frequency, or where the period $T > 1/\nu$, the time interval between successive collisions. Comparative d.c. and h.f. studies have confirmed this.

Energy losses and axial electric field

So far only the mean electron energy $\bar{\varepsilon} = kT_e$ has been calculated from the number loss of charge. To find the longitudinal field E in a uniform column the energy losses must be determined. This can be done either by integrating over the various types of losses and assuming, for example, a Maxwellian electron energy distribution, or by using the average loss factor κ per collision, which gives a lower boundary. Figure 7.17 shows $\bar{\kappa} = f(E/p)$ for different gases as obtained from electron swarm measurements of random and drift velocities $[v_d(v)^{1/2} = (\kappa/2)^{1/2}]$. At low E/p, $\bar{\kappa} = \kappa_{elast} = 2m/M_i$, which at higher E/p rises because of inelastic collisions. By balancing the energy per second an electron takes from the field E and that lost in elastic and inelastic collisions, one obtains with $\lambda_e = \lambda_1/p$, substituting for $v_d/\bar{v}$, and writing κ instead of $\bar{\kappa}$

$$eEv_d = \kappa\tfrac{3}{2}KT_e\,\bar{v}/\lambda_e \qquad (7.30)$$

$$E/p = \frac{(2\kappa)^{1/2}}{\lambda_1}\left(\frac{kT_e}{e}\right)\frac{3}{2} \qquad (7.31)$$

where kT_e/e is the volt-equivalent of T_e (1 eV = 11 600 K) and λ_1 is the averaged electron mean free path at 1 Torr, 273 K. For a positive column in neon ($V_i = 21\cdot6$ V, $\lambda_1 = 0\cdot1$ cm Torr^{-1}) with $R = 1$ cm, $p = 1$ Torr, $T_e = 35\,000$ K. From equation 7.31, $E/[p(2\kappa)^{1/2}]$ can be found by trial and error using figure 7.17 giving $E = 1\cdot5$ V cm^{-1}. This value is slightly smaller than the observed one since energy losses by metastables destroyed on the wall and radiation have been neglected.

We shall now briefly deal with losses caused by recombination, bremsstrahlung and cyclotron radiation, the latter arising from plasmas in magnetic fields. The non-uniform 'striated' column will not be considered here.

A free electron interacting with a quantum (photon) is scattered elastically with a cross-section of about $6\cdot7 \times 10^{-25}$ cm^2; yet it can abstract a larger fraction of the quantum energy if the equivalent wavelength of the quantum becomes of the order of Compton wavelength, i.e., $\lambda_0 \sim 2\cdot4 \times 10^{-2}$ m, corresponding to a quantum energy $h\nu \simeq 0\cdot5$ MeV. When an electron of low energy $\varepsilon_0 < eV_i$ recombines with an atomic ion, e.g., with H$^+$, to form atomic hydrogen in the ground state, a quantum $h\nu$ is emitted according to

$$H^+ + e \rightarrow H + h\nu$$

where $h\nu = eV_i + \varepsilon_0$, eV_i being the ionization energy (13·6 eV); if recombination into H* (stepwise) occurs, $h\nu = e(V_i - V^*) + \varepsilon_0$.

The recombination loss rate in J cm^{-3} s^{-1} is given by

$$(h\nu)\frac{dN_r}{dt} = (h\nu)\rho_e N^+ N_e = (h\nu)\overline{q_r v_e}N^+ N_e \qquad (7.32)$$

where ρ_e is the recombination coefficient in cm^3 s^{-1}, q_r is the cross-section and v_e

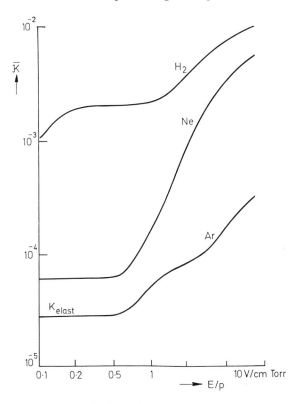

Figure 7.17. Mean loss factor $\bar{\kappa} = f(E/p)$ is various gases.

the electron energy. ρ_e rises as v_e is lowered because recombination is facilitated the smaller the relative velocity and hence the longer the 'contact-time' between electron and ion. Since for a head-on collision the linear momentum $h\nu/c$ of the emitted quantum is very small compared to mv_e, and since the angular momentum $p = mv_e r$, the incoming electron has to satisfy the selection rule, $p = h/2\pi \, \Delta j$, in order to have a reasonable chance of forming an atom. Only those electrons coming from certain directions with specific energies will qualify and hence electron–atom recombination coefficients are small (ρ_e order 10^{-12} cm^3 s^{-1}).

In highly ionized gases the position is more favourable, since for charge concentrations $> 10^{12}$ cm^{-3} 'two-electron recombination' into excited atoms frequently occurs. Figure 7.18 shows $\rho_e^* = f(N_e, T_e)$ for

$$H^+ + 2e \rightarrow H^* + e$$

and

$$He^+ + 2e \rightarrow He^* + e \quad \text{when } n > 5$$

resulting from calculations. The presence of the second electron thus helps to balance momenta and to remove the excess energy.

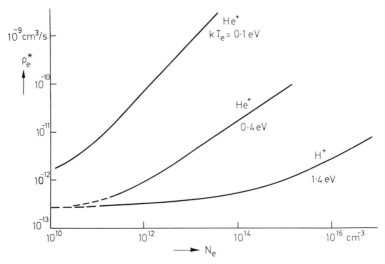

Figure 7.18. Two electron recombination coefficients ρ_e^* as a function of the electron concentration N_e and temperature kT_e. The process leads to excited atoms.
(D. R. Bates and A. E. Kingstone, 1961, *Nature* **189**, 652.)

Relatively large values of ρ_e are observed for electron–molecular ion recombination which occurs in rare gases at larger p and in ordinary molecular gases; for example

$$He_2^+ + e \rightarrow He^* + He$$

$$O_2^+ + e \rightarrow O^* + O^*$$

whereby the excess energy appears as kinetic energy of the resultant species. The last two reactions at $T_e = T^+ = T_{gas} = 300$ K give $\rho_e \sim 10^{-9}$ and 10^{-7} cm^3 s^{-1} respectively. ρ_e falls as T_e rises, as with atomic ions. He* and O* here have not been clearly identified. For H_2, O_2, and so on, $\rho_e \propto T_e^{-1/2}$ and $T^{-3/2}$ at lower and higher T_e respectively, while in rare gases $\rho_e \propto T_e^{-1/2}$ approximately.

In a fully ionized plasma in zero magnetic field the dominant loss is often the 'bremsstrahlung'. A free electron passing close to a positive ion (or an atom) is first accelerated and then retarded (scattered), a concept introduced in Chapter 4. As a result of this, in each event a fraction f of the kinetic energy of the electron is converted into a photon. Since $0 < f < 1$, the bremsstrahlung has a continuous spectrum, unlike atomic or characteristic X-ray radiation. The total intensity I_B (in W cm^{-3}) is proportional to the product of electron and ion concentration $N_e N^+$, to the electric field acting on the free electron ($\propto Z^2$) and to the relative velocity $\bar{v}_e \propto (T_e)^{1/2}$. Numerically the energy loss is for particles of atomic number Z

$$I_B = 1.5 \times 10^{-34} Z^2 N^+ N_e (T_e)^{1/2} \tag{7.33}$$

where the constants k, m and so on are included in the numerical factor. For a hot H^- plasma with $T_e = 10^8$ K, $N_e = N^+ = 10^{15}$ cm^{-3}, $I_B = 1·5$ W cm^{-3} and the mean wavelength of radiation corresponds to the interaction time ($=$ atomic diameter/$v_e \simeq 10^{-8}/10^{10}) = 10^{-18}$ s, i.e., to X-rays of the order of $0·1$ nm. All directions of emitted radiation are equally likely. In non-ideal plasmas radiation from impurity ions and from recombination of positive and negative ions can play an important role.

When the plasma is in a magnetic field B so that electrons with a component v_e perpendicular to B gyrate tightly around the lines of B (the Larmor radius of the ions is very much larger), then the radial acceleration of the electrons is accompanied by cyclotron radiation of angular frequencies (B_G in G, B in T)

$$\omega_c \left(= \frac{eB_G}{mc} \simeq 10^{11}B \right), 2\omega_c, 3\omega_c, \text{etc.} \tag{7.34}$$

In accordance with equation 4.25, the energy produced in W cm^{-3} is with ε, the electron energy in electron volt, and N_e in electrons cm^{-3} and B in gauss

$$W_{rad} = c_1 N_e B^2 \varepsilon \tag{7.35}$$

where $c_1 = 6·4 \times 10^{-28} \sin^2 \alpha$, α is the angle $(v_e B)$ and the direction of radiation is parallel to B. Since the frequencies ω_c are usually of the order 10^9 to 10^{10} Hz or more, the radiation is absorbed in the plasma because $\omega_e \sim \omega_p$, the electron plasma frequency. Thus radiation cannot escape and the losses are small. Yet when $T_e > 10^8$ K, the 'betatron' radiation losses become dominant. In this case $2\omega_c$, $3\omega_c$ and so on are contained to a large degree in the generated radiation, and absorption in the plasma ceases.

7.8. Quiescent plasmas

The drift motion of charges in any conductor is in general accompanied by electric disturbances caused by the random motion of the large number of electrons present. A sensitive electric field detector will therefore register a wide frequency band of these statistical fluctuations, often described as 'electric noise'. (Gauging the mean square intensity of the fluctuations is also a means of measuring the strength of the electric current.) There is a variety of other causes producing such fluctuations which originate from non-uniform regions of the electric plasmas; examples are the cathode–gas interface of a glow discharge, the boundary between the cathode dark space and the negative glow or the region adjacent to the anode. Also, these electrical fluctuations are, of course, accompanied by light fluctuations, which have been found to appear in various parts of the spectrum.

If the above assumed cold cathode of a glow discharge is replaced by a thermionically (electron) emitting surface, then the relative intensity of the disturbance is greatly reduced and the intensity distribution, i.e., the intensity versus frequency curve, changes its shape. The presence of positive ions makes itself felt in the noise and so does the gas temperature when the electron temperature of the plasma is not too different from the former and when the coupling between the two phases, the energy exchange between electrons and neutral gas, is strong. A low noise or quiescent plasma is of particular importance in the study of wave propagation of different kinds, because the signals picked up by electrostatic probes and other diagnostic tools, after a wave is launched, have to be discriminated from the earlier described 'background noise'. One type of a widely used quiescent plasma, the Q-machine plasma, will now be described.

Such a plasma is formed by mixing singly charged positive ions, usually alkalis, with electrons. Both can be emitted by the same electrode (figure 7.19). A tungsten plate C is kept by the filament F at about 2500 K and emits *in vacuo* a copious number of electrons of mean initial energy $\bar{\varepsilon}_0 \sim 0.25$ eV. Thus near C (in plane M), in the absence of positive ions, a potential minimum of order -2 V with respect to C develops. This allows only electrons with $\varepsilon > 2$ V to cross the barrier, whereas the rest, often the majority, are turned back towards C. Obviously M acts as a velocity selector, and the velocity component normal to C and the corresponding electron energy is the controlling parameter. The electrons crossing M are restrained from moving towards the wall of the adjacent tube by a uniform axial magnetic field B (~ 2 kG ~ 0.2 T). Consider now the change in M when a positive space charge is present.

The positive ions formed are the result of alkali atoms (Na), evaporated from a small oven, striking C. Since $\phi_W \sim 4.5$ V, and $V_i(\text{Na}) \sim 5.1$ V, the ratio of the numbers Na^+/Na leaving C is approximately $\exp[-(\phi - V_i)e/kT] \sim 0.1$. By placing a cooled electrically floating electrode D at the end of the long (cooled) metal tube, the vapour pressure is kept at $\sim 10^{-6}$ Torr and the density is 10^{10} atoms cm^{-3}. The ions of initial energy ~ 0.25 eV are accelerated towards M whose potential (because $\rho_{\text{net}} = \rho^+ - \rho_e$) has now been greatly reduced so that a larger number of electrons can pass M. The equilibrium value of V_M at M permits an appropriate number of electrons and ions to cross M, so that beyond M a quasi-neutral plasma is formed which fills a cylindrical volume of about 3 cm diameter at the centre of the long tube of about 15 cm diameter. This uniform plasma reaches equilibrium by virtue of frequent collective scattering collisions,

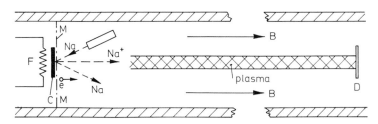

Figure 7.19. Q machine.
F = filament; $C = M - N$ plane of potential minimum; $C - D \simeq 0.8$ m; tube $\phi = 0.15$ m; Na plasma $\phi \simeq 2.5$ cm; $B = 0.2$ T; D = collector.

(see Chapter 5) between electrons and ions which are at different temperatures. The plasma maintains its uniformity, i.e., $N_e \simeq N^+$ over long distances although $T_e \neq T^+$.

Having discussed how a quiescent 'collisionless' plasma of charge density 10^7–10^8 cm^{-3} can be established, it should be briefly added that the negative glow and the positive column of a glow discharge (see Section 7.6) at $p \sim 0.1$–10 Torr represent gas plasmas with charge densities greater than 10^8 ions cm^{-3} in neutral gases at room temperature whereas at higher p the gas temperature is often considerably larger. Microwave and high frequency discharges are also used as plasma sources.

7.9. Plasma diagnostics

This fashionable heading covers the methods or principles of measurements of the main plasma parameters. Because of the extensive treatments available (see Further Reading) only a few methods are selected for discussion. The quantities and some of their distributions are as follows.

The concentration (number density) of neutral particles, excited particles, ions and electrons; the temperature of electrons and ions; the energy distribution of ions and electrons; the sheath thickness; the sheath field and the floating potential; and the local magnetic field.

Concentration of neutral particles

The gas density δ can be obtained from measuring p and T. p can be obtained absolutely by means of a McLeod gauge which—if fitted with capillary tubes of different sizes—can be used between 760 and 10^{-4} Torr, or down to 10^{-6} Torr. In the low range of p, 1–10^{-5} Torr, a Pirani gauge records continuous changes in p but gives a relative pressure value only. For plasmas which are found in a nearly fully ionized gas, the small concentration of neutrals and their nature can be determined by simple mass spectroscopy. The average gas temperature in a discharge tube can be approximately obtained by using a thermocouple in contact with the glass wall or by observing the change of gas pressure in a second vessel, both of known volume, connected to that with the discharge. The local gas temperature can be derived from interferometric measurements which give the gas density δ, related to the refractive index μ of the gas by $\delta \propto (\mu - 1)$. In cases of high T, δ and from it μ can be found from measurements of the velocity of sound, the range of α particles, the decay (scattering) of fast electrons and the absorption of X-rays of wavelength 0.1–1 nm. A 'correction' for dissociation has to be made in circumstances in which the ideal gas law is no longer applicable.

In highly ionized plasmas the density of neutral species is relatively low. It can be measured by passing line radiation through the plasma, which is modulated to separate it from the background, and determining the transmitted intensity which is weakened by the absorbing neutrals. This has been done, for example, in an H plasma where the absorption of the H_α (Balmer line 656·3 nm) gave a neutral concentration of 10^{10} cm^{-3} at charge concentrations of 10^{14} cm^{-3}.

Concentration of excited particles

The population density of short-lived ($<10^{-8}$ s) excited particles can be found spectroscopically from the intensity of radiation emitted, which is proportional to the particle concentration of the relevant species; the calibration requires an absolute measurement of energy associated with the line or band observed. The population density of long-lived ($>10^{-4}$ s) metastable particles can be found from the refractive index or by means of calorimetry or from electron emission. In the latter case the efficiency of energy exchange between the metastables and the calorimeter surface, or the probability of electron emission for a metastable incident on the emitting surface must be known. Chemical changes have also been used to measure the relative concentration of metastable atoms and molecules as well as photo-ionization whereby monochromatic light is absorbed and ions are produced. Very large excited populations have recently been found in alkali vapours (Carré 1981).

Concentration of ions

The absolute (positive) ion density can be obtained from measurements with a single Langmuir probe consisting of a wire which protrudes from a sealed thin glass tube with a flat end. The free length of the wire must be less than several tube diameters to avoid effects due to charges residing on the tube surface affecting the field of the wire. The variation of probe current i_p with the probe voltage V_p is found by changing V_B. For the range $|V_p| \leq |V_f|$, the 'floating potential' V_f corresponds to $i_p = 0$; $i_p = \int (V_p)$ is shown in figure 7.20. Note that at V_f, $i^+ = i_e$. As V_p is made increasingly negative relative to V_{pl}, the potential in the undisturbed plasma surrounding the probe, more positive ions are attracted and electrons (even the fast ones in the distribution) repelled, until under the simplest circumstances, all ions which have crossed r_s, the 'sheath boundary', (figure 7.21) are collected by the probe so that saturation is observed, i.e., i^+ is independent of V_p. In the majority of cases (dashed curve), owing to the variation of the 'collecting volume', $i^+ = f(V_p)$ tends to become saturated but rises slowly when V_p is further increased; at large V_p (at P_3) the probe becomes the cathode of an unwanted glow discharge.

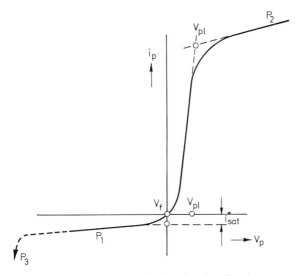

Figure 7.20. Semi-logarithmic Langmuir cylindrical probe plot of probe current i_p against probe voltage V_p (with respect to the anode).
V_f, V_{pl} = floating and plasma potential; i_{sat} = ion saturation current; P_1 = positive ion section; P_2 = section where $i_p = i_e$; P_3 = transition to a glow discharge.

The equivalent saturation current i_{sat} can be estimated as indicated in figure 7.20 and from

$$j_{sat}^+ = eN^+ v^+ = i_{sat}^+/A \propto \frac{V^{3/2}}{d_s^2} \qquad (7.36)$$

where V is the voltage drop across the sheath and $d_s = r_s - r_p$. N^+ in the plasma

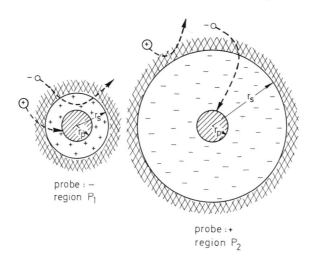

Figure 7.21. Sizes and charge trajectories of cylindrical probe sheaths.

can be found where $A = 2\pi r_p l$ is the probe area, l and r_p the length and radius of the wire, r_s the sheath radius and $v^+ \leq (kT_e/M^+)^{1/2}$, provided that no collisions occur in the sheath and its thickness is not more than the probe radius r_p. v^+ is the 'perpendicular' (often called Bohm) velocity at the sheath edge which is usually 10^5–10^6 cm s^{-1}. However, this is an over-simplified approach (see Further Reading). In ordinary (quasi-neutral) plasmas $N^+ \simeq N_e$ and thus another method of measuring N^+ is to find N_e, as shown below.

Concentration of electrons

The electron concentration N_e can be measured by, for example, irradiating the plasma with electromagnetic radiation approaching the angular frequency $\omega = \omega_{pl}$, the plasma frequency, at which nearly the full radiation is 'reflected' by dispersive effects. Since $\omega_{pl}^2 = e^2 N_e/m\varepsilon_0$, N_e can be determined. There are other, spectroscopic, methods all based on observing line widths from which N_e can be derived.

To find N_e in pulsed plasmas, laser interferometry is a useful tool. By counting the number of fringes from the moment the discharge starts, the change in average electron concentration can be found. The suitable laser wavelengths for plasmas with $N_e > 10^{10}$ cm^{-3} are in the red and near-infra-red. Probe measurements give N_e as well (see below; for fully ionized vapours see Further Reading).

Temperature of electrons

When a cylindrical probe (a thin short wire) immersed in a plasma is connected to a variable voltage V_B, the probe current i_p consisting of ions is relatively small, if the probe potential V_p is kept sufficiently negative relative to the plasma (figures 7.20, 7.21). As the probe potential V_p is raised, i.e., made more positive, the probe current decreases because more electrons of the distribution are collected by the electron-repelling probe, until at V_f, the floating potential, i_p becomes zero because i^+ and i_e cancel. For larger V_p, i_p not only changes its direction but a fast (exponential) rise is observed up to a 'knee' near P_2, figure 7.20; i_p to the right of V_f is thus essentially an electron current, the positive ions being now repelled. At $V = V_{pl}$, the plasma potential, which is a reference potential characterizing the undisturbed equipotential plasma, i_e displays a kink and then rises more slowly with V_p. This kink is, of course, seldom clearly pronounced, as will be shown later.

The section of the 'probe characteristic' between V_f and V_{pl} where $i_p \sim i_e$ is, if plotted as log $i_e = f(V_p)$, in most cases a straight line; since for a plane probe

$$j_e = eN_e(v_e/4) \exp(-eV_p/kT_e) \tag{7.37}$$

where $(v_e/4)=(kT_e/2\pi m)^{1/2}$; v_e is the average of the perpendicular velocity component, V_p is the voltage between the probe and V_{pl} the plasma potential. It is assumed that the potential difference between the anode and the plasma $(V_a - V_{pl})$ remains constant, independent of i_p.

The electron temperature is found from the slope of $\ln j_e$ versus V_p, which from equation 7.37 is

$$\frac{d(\ln j_e)}{dV_p} = \frac{d(\ln j_e)}{dV_B} = \frac{e}{kT_e} \tag{7.38}$$

kT_e/e is obtained by selecting two points on the ordinate, for example $2\cdot72\,(=|e|)$ apart and finding the corresponding difference ΔV on the abscissa (figure 7.22); 1 eV corresponds to 11 600 K. Note that V_p in equation 7.38 is a negative quantity, as seen in figure 7.22, while the plot makes $j_e > 0$. Equation 7.37 also reads $\ln j_e = \ln(eN_e v_e) - (eV_p/kT_e)$ where $v_e/4 = (kT_e/2\pi m)^{1/2}$. Since now j_e, T_e and V_p are known, N_e can be determined. For $V_p = V_{pl}$ at the imaginary kink near P_2 i.e., with the probe at plasma potential, the exponential term in equation 7.37 is unity and by substitution we find

$$(j_e)_{V_{pl}} = eN_e(kT_e(2\pi m)^{1/2} \tag{7.39}$$

With $j_e = 0\cdot7\ \text{A cm}^{-2}$, $v_e/4 = 2\cdot2 \times 10^7\ \text{cm s}^{-1}$ for $kT_e = 1\cdot7\ \text{eV}$ we obtain $N_e = 1\cdot9 \times 10^{11}$ electrons cm^{-3}. From the observed ion saturation current density (figure 7.22) $j^+ \simeq 0\cdot7 \times 10^{-2}\ \text{A cm}^{-2}$ and $v^+/4 = 5 \times 10^4\ \text{cm s}^{-1}$, obtained from $2\cdot2 \times 10^7 (m/M)^{1/2} \simeq 2\cdot2 \times 10^7\ (1/200)$, one finds $N^+ \simeq 8 \times 10^{11}$ ions cm^{-3} or about four times the value for N_e. However, if we use the mass and temperature ratios, and apply $v^+ \simeq (kT_e/M)^{1/2} = 2\cdot8 \times 10^5\ \text{cm s}^{-1}$, we find $N^+ \simeq 1\cdot4 \times 10^{11}$ ions cm^{-3}, in reasonable agreement with $N_e = 1\cdot9 \times 10^{11}$ electrons cm^{-3}. This shows that the acceleration of ions by the probe field in the originally assumed undisturbed plasma cannot be neglected.

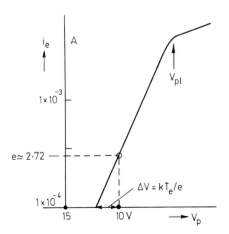

Figure 7.22. Evaluation of T_e in eV from the curve in figure 7.20 (in Ne).

Let us turn our attention to the positive space charge sheath surrounding the cylindrical probe. If the plasma density is high this layer is essentially thin, compared with r_p (figure 7.21), since for a negative probe the Debye length Λ_p is relatively small. Equation 7.36 suggests a thickness d_s of the order 1 mm for $N^+ \sim 10^{10}$ ions cm^{-3}. For $j^+ = j^+_{sat}$, $d_s \propto V_p^{3/4}$. As V_p is raised up to V_f, d_s increases because the larger V_p is the nearer the probe is to V_{pl}; at this point the probe wire represents a conducting surface on which charges are neutralized but at which no field lines emerge or end, and d_s becomes indefinite and disappears. For positive probes, $V_p > V_{pl}$ becomes very large (many cm) and the sheath is electronic. The sheath concept is only useful if a boundary between it and the undisturbed plasma can be defined. That has been virtually abandoned by assuming a finite ion velocity v^+ (see above), which depends on kT_e and M, but not on V_p. v^+ results from the penetration of the probe field into the plasma region beyond the fuzzy 'boundary'. It follows that the undisturbed plasma is separated from the sheath by a 'pre-sheath' in which the ions no longer have a pure Maxwellian distribution and in which a corresponding potential drop has to be allowed for.

Temperature of ions

There are few direct methods of determining T^+. Its value can be derived from the measured ion energy distribution using a retardation method, an ion analyser or a spectrograph (see below). Recently, T^+ has been measured with a thin cylindrical plus a large spherical probe: the former receives $i_{sat} \propto (N^+ T_e/T^+)$, so if $N^+ = N_e$ and T_e are known, T^+ can be found. The accuracy is, however, somewhat restricted. $T^+ = f(p)$ at low pressure is shown in figure 7.31 (p. 162).

Energy distribution of ions

The ion energy or velocity in a specified direction can be found by means of an analyser which measures the relative number of ions dN^+/N^+, usually of given charge to mass ratio, as a function of their energy ε in a small energy range $d\varepsilon$. The ions are supposed to originate in a gas-filled space, whereas the analyser is highly evacuated; this fact restricts the flexibility of the method. Moreover, at large p, because of frequent collisions with neutrals, T^+ is not much above T_{gas} and hence the accuracy of measurement is low. In general, isotropy is then complete, i.e., the velocity distribution is independent of direction, unless large electric fields or local or temporal magnetic field changes act on the ions. Up to now the ions have been assumed to be in their spectroscopic ground state.

Excited ions, however, emit 'spark lines' and, if in equilibrium with the unexcited ions, make it possible to apply a spectroscopic method of energy analysis. An excited particle moving with velocity v towards an observer at an angle θ with respect to the line of sight, while emitting light of wavelength λ, appears to change its wavelength by $\Delta\lambda$ so that according to Doppler

$$\Delta\lambda/\lambda = (v/c)\cos\theta \qquad (7.40)$$

In the case of a distribution of identical excited ions, the Doppler shift $\Delta\lambda$ produces a band. The intensity in the band is a measure of the number of ions while $\Delta\lambda$ gives the velocity and energy. For example, the H_α line (656·3 nm) of width of order 0·01 nm is displaced by $>0·5$ nm if emitted by a 1 keV ion. The light emitted by H* is the result of a charge transfer collision with H (or H_2)

$$H^+ + H \rightarrow (H_2^+)^* \rightarrow H_{slow}^+ + H_{fast}^* - \Delta$$

where Δ is a small amount of energy (here 10 eV) necessary to excite H. Fast positive ions in their ground state may induce light emission by charge transfer collisions

$$\text{ion}_{fast} + \text{atom}_{slow} \rightarrow \text{atom}_{fast}^* + \text{ion}_{slow}$$

The fast excited atom acquires nearly the full kinetic energy K of the former ion, namely $(K - eV^*)$. Again from the intensity distribution in the Doppler-shifted band the ion energy distribution $f(\varepsilon^+)$ can be found.

Energy distribution of electrons

This can be determined by the Langmuir–Druyvesteyn method. If the distribution $f(\varepsilon)$ is isotropic, it can be derived from a single-probe curve $i_p = f(V_p)$ if $i_p = i_e$, that is, approximately between V_f and V_{pl}, from

$$f(\varepsilon) = C|V_p^{1/2}|\frac{d^2 i_e}{dV_p^2} \qquad (7.41)$$

where $f(\varepsilon)$ is the relative number of electrons per unit volume in the range $\varepsilon{-}\varepsilon + d\varepsilon$ and C is a constant. Since graphic double differentiation is not accurate, various electric methods such as superposition of small a.c. components or modulated voltages are used (see Swift and Schwar, 1970). It can be shown that analysis of a 'double probe' curve gives the energy distribution in the tail.

As with positive ions, a highly evacuated electron energy analyser, attached to a discharge tube, can extract a "beam or swarm sample" which gives a distribution corresponding to the energies of the electron gas near the wall. In order to analyse the electron energies normal to the anode surface of a low-pressure glow discharge, one can extract the electrons through a small gap or hole in the anode which leads to a space with a uniform well-shielded magnetic field behind it. The electrons of different velocity are forced to travel *in vacuo* along a semi-circular path and then impinge on, for example, a photographic plate. The degree of blackening (the photographic density) is a measure of the relative number per unit energy range $d\varepsilon$ while the radius of their path is a measure of ε. Instead of a photographic plate, a collector electrode at variable retarding potential V placed behind a small hole in the anode registers a current i which, if differentiated once, gives $f(\varepsilon) = V^{1/2} \, di/dV$, the energy distribution in one direction.

The use of two equal adequately separated probes to find i_{sat}^+, N^+ and kT_e is often advantageous. Such a 'double probe' gives a symmetrical curve (figure 7.23), from which one obtains

$$kT_e/e = \Delta/4 \qquad (7.42)$$

where Δ is the potential difference between the two 'kinks'. The floating (well isolated) measuring circuit makes a reference potential unnecessary. However, compared with the single probe, only a short section of the electron distribution comprising the fast electrons is used in this method. N^+ can be found as shown above in the subsection on 'Temperature of ions'.

Sheath thickness, field and potential of an isolated plane probe

The electric field E near a large 'floating' plane metal electrode in a low-pressure Ar plasma has been measured by observing the deflection of an electron beam as a function of the distance y from the electrode (figure 7.24). By integration of $E(y)$, after extrapolating it to $y = 0$, $V_f = 20$ V is obtained; this value is compared with the measurement yielding $V_f \sim 20$ V and with theory. Data are given in the figure's legend.

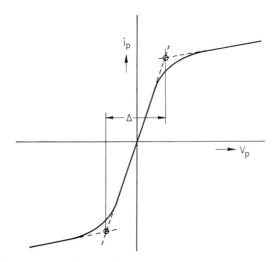

Figure 7.23. Double probe curve (linear plot).
T_e in K $= 1 \cdot 17 \times 10^4 \times \Delta/4$, Δ in V.

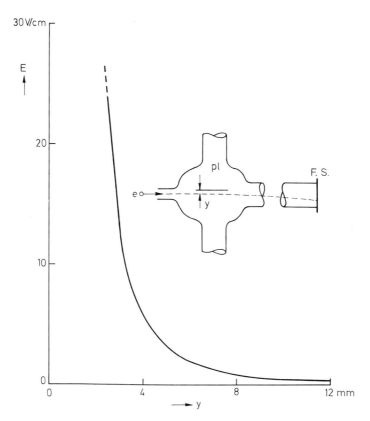

Figure 7.24. Electric field E near a plane floating electrode in an Ar plasma at 0.85 mTorr, agreeing with free-fall theory.
(Goldan 1970.)
e = electron beam; F.S. = fluorescent screen; pl = plasma.

From theory, with $v_e = (8kT_e/\pi m)^{1/2}$, $v^+ = (kT_e/M)^{1/2}$, $M/m \sim 8 \times 10^4$, $kT_e/e \sim 3$–5 V, the floating potential

$$V_f = \frac{kT_e}{e} \ln \left(\frac{N_e v_e}{N^+ v^+} \right)^{1/2} = \frac{kT_e}{2e} \ln \left(\frac{8M}{\pi m} \right) = 1.6 \times 12.2 \simeq 19 \text{ V}$$

Other values observed were: for $T_e \sim 4.4 \times 10^4$ K, $N_e \sim 9 \times 10^8$ cm^{-3} at $p \sim 10^{-3}$ Torr in Ar, $\Lambda_D \sim 7(T_e/N_e)^{1/2} \sim 5$ mm.

Local magnetic fields

The simplest magnetic probe is a coil of effective area a with n windings which are threaded by the magnetic flux of the field B. Its value, averaged over a, is found, for example, oscillographically, by observing the voltage e induced in the coil. Since $|e| = an\,dB/dt$, its time integral is proportional to $B - B_0$. With small coils ($a \sim 0.2\ cm^2$), a time resolution of 10–100 ns has been obtained.

The local field B can also be measured by using the Hall effect in semiconductors such as indium arsenide (or indium antimonide), which produces a relatively large voltage e across a specimen when a current i is passed along it while B acts perpendicular to i. Since e appears because positive holes are deflected by B to one side of the semiconductor, the delay times are very short compared with the time resolution of the oscillograph. Thus $e(t)$ is a direct measure of $B(t)$. Note that the Hall coefficient is temperature-dependent.

Another of the numerous methods for measuring strong local fields in a plasma, without causing disturbances, uses the Faraday effect. Here the rotation of the plane of polarization is observed when an infra-red or microwave beam is passed through a plasma. The rotation α_F, in degrees per cm length of plasma, is

$$\alpha_F = 10^{-23}\lambda^2 N_e B_G \tag{7.43}$$

with λ in μm, N_e in cm^{-3} and B_G in gauss (10^{-4} T). Thus when λ and N_e are known, B can be found. Since the linearly polarized electromagnetic wave, constituting the beam, interacts with the free electrons in the plasma, it follows that the rotation of the plane of polarization is caused by B which forces the electrons with v components perpendicular to B to move along circular paths (Larmor = cyclotron frequency ω_c) between successive scattering collisions. For $N_e = 10^{16}\ cm^{-3}$, $B_G = 10^3$ G, $\lambda = 10\ \mu m$, $\alpha \sim 10^{-2}$ degrees cm^{-1}. Though raising λ raises the accuracy, this is accompanied by a reduction in spatial resolution.

7.10. Anomalous plasma resistivity

When a fully ionized collisionless plasma of the type described in Section 7.8 is subjected to a radio frequency field E of constant value, it is found (a) that its resistance R' per unit length of plasma column decreases with increasing average electron concentration N_e as shown in figure 7.25 for three different gases (obviously, for a given value of N_e, R' is the smaller the larger the atomic mass M); and (b) that if the electron concentration N_e is kept constant and the field E is increased then the resistivity of the plasma, ρ_p (figure 7.26) is found to increase abruptly when E exceeds a critical value E_c. The changes (figures 7.25 and 7.26) occur in such a manner that they are not derivable from the classical relation $\rho = E/j = E/(eN_e v_d)$; the latter suggests that for constant E, $\rho_p \propto 1/N_e$, whereas, for constant N_e, ρ_p should rise steadily with E, conclusions which differ from both (a) and (b).

Therefore these effects have been termed 'anomalous resistivity'. The relation $\rho_p \propto N_e^{-1}$ above holds only for electrons drifting in weakly ionized gases, in contrast to the strongly turbulent (pulsed) and highly ionized plasma column

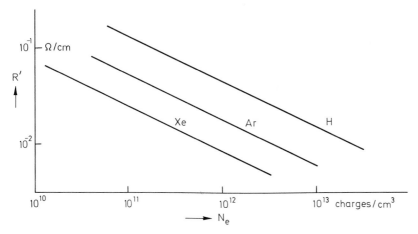

Figure 7.25. High frequency resistance R' per unit length of a plasma column $= f(N_e)$ in H_2, Ar and Xe for $E = 300$ V cm^{-1}. $R' = \rho_p/a^2\pi$.
(S. Hamberger and M. Friedman, 1968, *Physical Review Letters* **21**, 674.)
$a =$ column radius; $\rho_p^{-1} = 4\pi\varepsilon_0\omega_{pe}(M/m)^{1/3}$ where $\varepsilon_0 \sim 10^{-11}$ F m^{-1}.

(figure 7.25), showing that here collective electron scattering is predominant whereby $\rho_p \propto N_e^{-1/2}$. Thus R' depends less strongly on N_e and also varies with M^+, and hence $(M^+/m)^{-1/3}$, and this gives approximately the slope and displacement of the curves respectively. The sudden increase in ρ_p has been explained by the appearance of turbulent electric fields in the plasma which are excited when E_c exceeds a critical (oscillatory) velocity. We note that in a pulsed

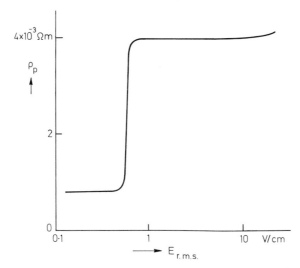

Figure 7.26. Anomalous plasma resistivity $\rho_p = f(E)$. For $E_{rms} > 0.6$ V cm^{-1} in H_2 turbulence (discontinuity) results.
(N. Demidov *et al*, 1967, *Soviet Physics, Doklady* **12**, 467.)
$E =$ pulse superimposed on damped wave when $N_e = 10^{12}$ cm^{-3}.

plasma where $T_e > T_i$ with T_e of the order 2 eV, in the absence of binary gas collisions, electron–ion collisions determine the electron drift velocity. The bulk of the electrons, accelerated by a r.f. field of $E > 0.4 \text{ V cm}^{-1}$, experience a retardation in the presence of these waves and this is thought to be the reason for the sudden increase in resistivity when E exceeds E_c. However, details of these processes are not clear. The anomaly seems to occur in turbulent plasmas only, i.e., when collective interactions make themselves felt.

7.11. Arc discharges

When the current in a glow discharge between cold metal electrodes in a gas at a pressure of 0.1–1 atm is gradually increased, the voltage across the discharge is seen first to rise (figure 7.27). However, when the current exceeds a certain magnitude, the voltage suddenly drops to about $1/10$ or less of its former value and an arc forms. Since there is usually a constant resistance in series with the discharge, the sudden drop in voltage is accompanied by a sudden rise in current. The luminous area at the cathode surface contracts to a small patch—the cathode spot. Spots of high temperature develop at the cathode and anode and in the spot areas the metal electrodes are ablated ('evaporated'). In general, the cathode vapour determines the colour of the emitted light. At the same time the gas in the discharge column acquires a very high temperature. The phenomenon describes the transition from a glow to an arc discharge when metal electrodes like copper or iron are used in a gas such as nitrogen or air. The above applies only to one kind of electric arc, the vapour arc. An example of the other type is the carbon,

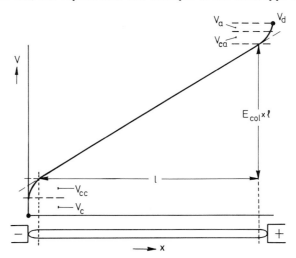

Figure 7.27. Potential distribution $V = f(x)$ of an arc with the transition (contraction) regions V_{cc} and V_{ca}, and the cathode and anode falls V_c and V_a.

silicon or tungsten arc, the latter often kept in nitrogen to avoid oxidation of the electrode material.

It is convenient to distinguish first between these two arc types since they differ essentially in the mechanism of emission of electrons from the cathode. When the temperature T_c of the cathode spot of a carbon arc in atmospheric air is measured, it is found that at its value of 3600 K the number of electrons emitted per unit area of the spot corresponds to electron current density j_c as given by Richardson's equation of thermionic emission

$$j_c = A_c T_c^2 \exp\left(-e\phi/kT_c\right) \qquad (7.44)$$

where $A_c \simeq 50$ A cm^{-2} K^{-2} and the work function of C, $\phi = 4\cdot4$ V.

Under these conditions electron emission occurs in the same way as is known from vacuum tubes with pure tungsten emitters. However, their filament is supplied with power from an external source, whereas in a self-sustaining arc the power to heat the cathode spot is supplied by the discharge; in fact, positive ions from the arc column pick up energy in the cathode fall and strike the cathode as well as slow and fast neutrals (metastables, charge transfer products) and deliver their energy to the cathode spot.

For a given arc current the energy drawn from the discharge manifests itself in the cathode fall in potential. Its value depends of course on other losses, such as heat conduction from the spot into the electrode, evaporation and light emission, as well as on the energy gain due to neutralization of the ion charge. Recent observations of cathode falls of arcs at atmospheric pressure for different substances are given in table 7.2 together with the temperatures of the arc spots.

In contrast to the carbon arc are ordinary vapour arcs. The temperature of their spots is relatively low (2000 to 500 K) and thus thermionic emission is negligible. Field emission accompanied by evaporation could account perhaps for large current densities but not for cathode falls of less than 10 V as are known to occur in mercury arcs (7–8 V, i.e., considerably below ionization potential). These arc spots move in general quickly over the cathode surface and are accompanied by a jet of luminous metal vapour. The spot size is order of magnitudes smaller and the current density larger than in a carbon arc (table 7.2), and although the

Table 7.2. Cathode and anode spot temperatures T_c, T_a, gas positive column temperature T_g, current densities j_c, j_a, and cathode and anode falls V_c, V_a, in arc discharges.

Cathode–Anode	Gas	T_c(K)	T_a(K)	T_g(K)	j_c (A cm^{-2})	j_a (A cm^{-2})	V_c (V)	V_a (V)
C–C	atm. air	3500	4200	6000	470	65	~4	~11
(5–10 A)	,, N$_2$	3500	4000	6000	500	70		
Cu–Cu	,, air	<2200	~2400	4700	10^5–10^6	10^3	16	10
(2 A)	,, N$_2$			5200	high			
Zn–C	,, Ar						11	4·5
(10 A)								
Hg–C	Hg	600	~300	—	10^5	—	8	~2
(10 A)	(low p)							

cathode fall is of the same order, these facts seem to indicate that the mechanism of the emission of the electrons is fundamentally different. It can be assumed that excited atoms striking the cathode convey to it enough energy to release electrons from the metal, while the accompanying positive ions keep the net space charge near zero level which accounts for the low cathode fall in potential.

The potential distribution $V(x)$ along the arc, x being the distance from the cathode, is shown in figure 7.27. There is a steep potential rise near the cathode corresponding to V_c that is followed by a curved section which goes over into straight line, the latter indicating the constant field in the positive column. Similarly, a curved section followed by a smaller steep voltage rise, V_a the anode fall, is found at the anode. The curved parts of $V(x)$ near the electrodes are due to the radially contracted parts of the column whose ends join the cathode and anode spot respectively. One can argue that since larger axial electric fields exist in regions in which the positive column contracts and since the current is the same throughout the discharge, larger losses will arise in these regions. This indeed is the case. When the positive column joins either electrode, in addition to radial losses in the positive column, axial thermal energy losses arise which cause the increase in the local axial field. The magnitude of the axial losses is considerable, because the electrodes forming the boundary of the column are solid thermal conductors.

The total voltage V_d across a moderately long arc (figure 7.27) is the sum of anode (V_a) and cathode fall (V_c), the voltage across two contracted regions (V_{cont}) and the positive column, assumed to be straight, uniform and thus with an electric

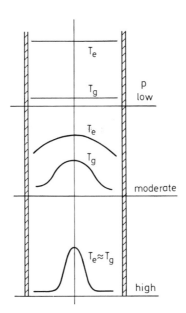

Figure 7.28. Radial distribuition of T_e and T_g in an arc column at different gas pressures.

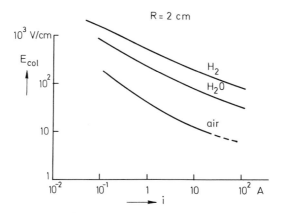

Figure 7.29. Electric field E_{col} in the arc column $= f(i)$ for different gases at 1 atm (slow gas flow).

field E_{col} independent of x, as in a glow discharge; but at larger pressures the column acquires a high temperature (figure 7.28) with a bell-shaped radial profile. Hence, if l is the length of the uniform part of the column

$$V_d = V_c + V_a + V_{cont} + V_{cont} + E_{col} l \qquad (7.45)$$

Only E_{col} will be considered in what follows and figure 7.29 shows its dependence on the current i for various gases at 1 atm. In order to obtain a long straight stable arc column (>5 cm) a gas flow is set up along a helical path around the positive column, or a circular component of flow is used by spinning the tube about its axis so that the colder dense gas is centrifugally moved towards the glass wall and the hot gas to the axis. Any temporary displacement of the hot arc column from the axis (figure 7.30) is counter-balanced by denser gas which is wedged in between the tube wall and the deflected column. Figure 7.29 also shows that the axial field E_{col} is larger in H_2 or H_2O than in the heavier molecular gases. The reason is that $E_{col} i$, the power loss per unit axial length of column, is large in gases of large 'effective' thermal conductivity such as H_2 and H_2O, as a result of effects involving dissociation. In principle, E_{col} can be numerically estimated from atomic data and the Lindemann–Saha equation (equation 7.46), which describes thermal ionization. Figure 7.31 is self-explanatory.

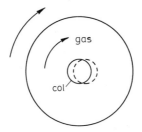

Figure 7.30. Arc column stabilization by rotating the surrounding (glass) cylinder.

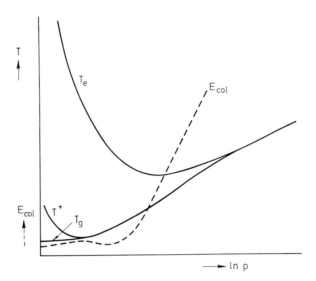

Figure 7.31. T_e, T^+, $T_g = f(\ln p)$ between 10^{-2} and 10^3 Torr.
$T_{e\,min}$ at $p \sim 10^2$ Torr; $N_g \sim 10^{18}$ cm^{-3}.

Applying this relation to the central part of the N_2 column and assuming there is an average gas temperature T of about $5–10 \times 10^3$ K with a V_i of about 15 V and the thermodynamic relation reads for $x_i \ll 1$ (low degree of ionization)

$$x_i^2 p \simeq (g_i/g_0)5 \times 10^{-4} T^{5/2} \exp(-eV_i/kT) \qquad (7.46)$$

where $x_i = N_e/N_0$ is the degree of ionization, p the actual gas pressure in Torr and N_0 the initial concentration of molecules. With $p = 760$ Torr, $N_0 \sim 3 \times 10^{19}$ cm^{-3}, $g_i/g_0 = $ order 1, $N_e = f(T)$ can be determined. Since the current

$$i = R^2 \pi e N_e v_d = f(R, T, \Lambda) \qquad (7.47)$$

where Λ is the effective thermal conductivity (i.e., including dissociation), R is the core radius and v_d depends on E/N with $N/N_0 \simeq 1/20 (= 300/6000)$, we find from equation 7.46 that $x \simeq 1 \times 10^{-7}$ and since per cm length of the column

$$E_{col} i \simeq \Lambda T \qquad (7.48)$$

with $\Lambda = 2 \times 10^{-2}$ W K^{-1}, then $R = 0.5$ cm, $N_e = 4 \times 10^{12}$ cm^{-3}, $j_e \sim 10$ A cm^{-2}, $v_d \sim 2 \times 10^6$ cm s^{-1}, and $E/N_0 \sim 4$, values which are in satisfactory agreement with observations. It is clear that T decreases as the column approaches the electrodes and if cooler gas only were present, x_i and N_e would drop. However, near the electrodes metal vapour with V_i of about 7–8 V (instead of about 15 V) is dominant and thus N_e is large, though T is lower than in the uniform column. The vapour and current density (j_e) rises gradually towards the electrode spots and the column contracts.

An interesting observation seems to support the view that from the

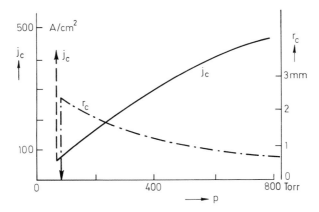

Figure 7.32. Cathode current density $j_c = f(p)$ of a carbon arc (i order 10 A) and the contraction of its spot at a critical pressure, p_c (the vertical lines of j_c and r_c have the same p value, p_c, but are separated to avoid confusion).

cathode spot of a vapour arc electrons are emitted by excited atoms available in copious numbers in a thin layer adjacent to the cathode and compressed by the influx of positive ions. The radius of this spot is very small ($\sim 10^{-7}$ m); the spot is a fast-moving and unstable entity and the current density is very large ($j_e \sim 10^5$ A cm^{-2}); the cathode spot radius on carbon at 1 atm is comparatively large, say 1 mm at 10 A, and the current density is only about 500 A cm^{-2}. However, one finds that the spot size r_c on carbon increases as p is reduced (figure 7.32) but at a critical value p_c the spot suddenly contracts and begins to move fast and erratically over the cathode surface (vertical lines), thereby producing black deposits on the inner surface of the evacuated vessel. Obviously the carbon arc cathode is 'ablated' at a relatively high rate. Therefore when p is low the emission seems to be no longer thermionic but that of the vapour arc. The transition from one to the other form has also been found with other substances. Though the current density j_c in the spot changes on transition by a factor of about 10^3, the change in cathode fall is very small (order 1 V).

The cathode fall of a thermionic arc can be further reduced when the cathode, instead of being heated from the discharge, is heated from an outside source. The total discharge voltage of, for example, an arc in argon at 1–10 Torr between an oxide-coated cathode and a cold anode about 1 cm apart can be as low as 5–7 V, but rises when the heating power and the cathode temperature are reduced. The rise in discharge voltage is expected to occur when the vacuum 'emission' current from the hot cathode becomes smaller than the arc current.

Electrodeless arcs with ring-shaped positive columns can be produced in any gas or gas mixture at low and high p by high frequency excitation using short cylindrical coils which induce eddy currents in the hot gas. Such arrangements are of interest to plasma chemists.

Mixed forms of glow and arc discharges are often observed. An example illustrates these hybrids. In Hg vapour we find small mobile arc spots on liquid

Hg cathodes to which either a diffuse, glow-like positive column or a constricted arc-like column is attached. Also known are cathodes with glows covering the whole Hg surface with either a diffuse or a narrow constricted column joined to it. Which of the four forms develops depends on i, p and R.

miscellaneous effects involving discharges and plasmas

To restrict the size of this section I shall only summarize the various manifestations and uses of electric discharges and plasmas. Though the scientific content in what follows will be condensed to a bare minimum, it may give the reader new ideas about possible future developments, in a technological sense as well as conceptually or mechanistically; with luck, he or she may develop a 'mental model'. If that helps in the comprehension of a proposed mechanism or the analysis of a 'novel' phenomenon, then the object of these notes has been achieved.

8.1. Gas explosions initiated by micro-sparks

The cause of tanker explosions, for example during washing operations or due to splashing in partially filled tanks in heavy seas, was finally traced back to electrostatic charges on droplets and in mist. Charges swept towards a metal surface gave rise to electric sparks in the existing inflammable atmosphere, particularly near projections. Recent research on charge generation accompanying the formation of spray or the disruption of ballast water surfaces has provided quantitative information on the minimum spark energy causing explosions, on the effect of protrusions, or spark-quenching and other factors controlling the ignition by sparks. Note that in this case the discharge is not actuated by an 'applied' voltage between two electrodes but by electric charges carried by small drops. Detailed studies of the fundamental processes have led to the development of practical means of preventing further loss of life and material.

8.2. Protection against lightning

Objects such as space vehicles, aircraft, trains, towers, chemical factories, storage containers and chimneys are all exposed to lightning, and recent research indicates how the risks of electric shocks, gas explosions, shock waves and thermal effects can be reduced. A lightning stroke between cloud and ground which is accompanied by intense light emission and chemical changes in the air is the final phase of a gas discharge, the so-called return stroke, whereby the bottom of the cloud is negative with respect to the ground. The maximum current in the return (leader) stroke is up to several hundred thousand ampere and is arc-like in appearance.

Its root (anode spot) leaves marks on the object hit. The return stroke is preceded by a multi-branched leader stroke of relatively low current. This precursor determines the final path and the end point of the return stroke which develops from the ground to the cloud.

Therefore, if the likelihood of an object on the ground becoming the anode of the return stroke depends on the electric field topology, i.e., the field pattern at and around the exposed object, its average value can be reduced by providing a positive space charge which surrounds the positive metal object to be protected. This space charge can be set up by the passing thundercloud field itself, which causes a spacious glow or a corona discharge to develop at and around an appropriate metal structure. The corona current in such structures is up to about 10^{-2} A; the discharge is established within several seconds, extending over a distance of 50–100 m from the structure. The resulting electric field is thereby reduced to less than the breakdown field in air ($30 \, \text{kV cm}^{-1} = 3 \, \text{V m}^{-1}$) by an average positive charge density of about $10^6 \, \text{C m}^{-3}$ or $10^7 \, \text{ions cm}^{-3}$. As a result of the artificial pre-ionization of a large volume of air, the development of a positive leader is impeded if not prevented. Since the pre-ionizing time is of the order of seconds and the speed of a commercial aircraft is about 600 miles per hour, or $300 \, \text{m s}^{-1}$, it follows that a protective ion cloud cannot be formed sufficiently fast around an aircraft.

8.3. Corona discharges for paint spraying and xerographic copying

Electrocoating involves a transfer of liquid droplets, dry powder particles and short fibres from a gun or sprayer onto a collecting earthed surface. For spraying liquid paint a high negative voltage (with respect to earth) is applied to a point electrode of a spray gun which produces a corona discharge in air. The paint, in the form of mist particles, originally carrying zero charge, acquires a relatively large negative charge, owing to ionization of air by the corona, and is driven away from the negative point towards the positive electrode, the earthed surface. The particle paths follow essentially the field lines, particularly in regions not too near to the corona point. The positive ions mainly produced around the point move only a short distance, take few positively charged mist particles with them and are neutralized on arrival at the corona point. In this way liquid paint can be transferred from a spray gun (atomizer) and uniformly distributed over a large area.

Another application of an electric discharge is the copying of typed documents, manuscripts or drawings by the xerographic method. It makes use of the corona discharge around a long thin rod in air and the photoelectric effect. The working principle of the Xerox machine is as follows: a corona deposits positive ions on a selenium-surfaced, insulated drum in darkness. Then an image of the strongly ultraviolet illuminated original (to be copied) is focused onto the drum which releases photoelectrons from the lit selenium sites but not from the dark areas. Since the electrons locally neutralize the ions, the lit drum areas are now neutral but the dark patches retain a positive charge. The drum is then covered with a fine black powder ('toner') which sticks to the charged sites. Thus when the drum is brought in contact with a pre-charged special paper it attracts the toner away from the drum. After being exposed to light to remove the charges, the paper with the attached toner passes a hot (fusing) roller which ejects the copy at a rate of up to one per second.

8.4. Corona discharges for separating particles

The separation of conducting from insulating particles in mining is achieved by dropping the mixture of mineral powders through a funnel onto the highest point of a rotating earthed steel drum (figure 8.1). The thin dust layer, on its motion downwards, is exposed to a flux of ions. These ions drift from the corona discharge that surrounds a wire (kept at a high voltage) which is fixed above and parallel to the axis of the drum. Ions of like sign convey their charge to all particles on the surface of the dust layer. Since the fraction conducting and in contact with the drum is discharged thus to earth, no electric force keeps it to

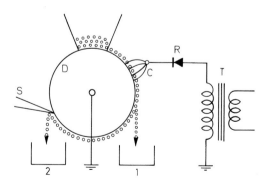

Figure 8.1. Separating insulating from conducting particles by a glow (corona) discharge C in air with a rotating drum D and collecting them in box 2 and 1 respectively.
R and T = rectifier and transformer respectively.

the surface of the drum, so it falls, due to gravity, into a compartment below. The insulating particles, however, which are charged up by the ions, experience an electric force which keeps them stuck to the drum. These particles are subsequently removed by a scraper, during their upward journey, and fall into a second compartment under the drum. Similar electrostatic methods are applied to separate the kernels of nuts from their shells (insulators), and garlic seeds from wheat (insulators).

8.5. Isotope separation by lasers

Turning to more sophisticated applications, the use of discharge lasers for separating isotopes which are required as fissionable materials in nuclear power stations is of interest. To separate, for example, the small amount of ^{235}U contained in natural uranium, mainly ^{238}U, one can irradiate the U vapour with laser light 1 which excites the isotope ^{235}U to a long-lived state which is photo-ionized by a laser 2 according to

$$^{238}U + {}^{235}U + h\nu_1 \rightarrow {}^{238}U + {}^{235}U^*$$

$$^{238}U + {}^{235}U^* + h\nu_2 \rightarrow {}^{238}U + {}^{235}U^{(+)}$$

The partially ionized products are passed through an electromagnetic separator and collected. Though this process would hardly be of practical interest, the method illustrates one principle of irradiative isotope separation.

8.6. Gas dynamic high-voltage generation

The conventional electrostatic van de Graaff generator producing high voltages up to about 10 MeV and small currents makes use of an insulating conveyor belt onto which electric charges are sprayed at earth potential by a corona discharge. These charges are then moved mechanically to a distant high voltage electrode in a direction opposite to the electric field of the latter. On arrival at the high voltage electrode the charges on the belt are at a potential somewhat larger than that of the high voltage electrode, so that a corona discharge between the charged belt and the electrode develops which removes them to the electrode.

A modern variant of this method is the electro-gasdynamic generation of high voltages. Here the charged belt is replaced by a gas jet which blows charged dust particles to the high voltage electrode. Hence moving mechanical parts are replaced by moving gases, the electrical charges are again produced by a corona but the ducts for guiding the charged dust particles present new technical problems.

8.7. Electric precipitation

The electric precipitation of fog, dust and other particles in air can be demonstrated by placing a thin wire at the axis of a cylindrical conductor and applying to it a high, usually negative, potential. The resulting corona forms around the wire charges and sweeps the charged particles that are being carried by the air flow towards the cylindrical electrode. The unit length current i' observed as a function of the applied potential V is shown in figure 8.2, (a) for clear air (high small-ion mobility) and (b) for dusty air (reduced average mobility). Below the corona onset potential V_0 no discharge develops. At $V > V_0$ a relation $i' = f(V - V_0)$ applies, implying a constant electric field at the wire's surface.

The physical reasons for the appearance of a spark discharge in atmospheric

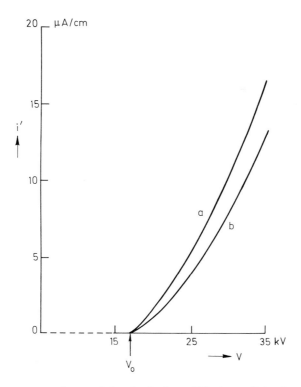

Figure 8.2. Corona current i' per unit length of wire $= f(V)$, the applied voltage V_0 being the starting voltage.
Curve $a =$ air, curve $b =$ air with dust.

air between two parallel plates and of a corona discharge between a wire in a cylinder (or a point electrode and a plane) is obviously the difference in field configuration. A uniform field thus leads to a sudden breakdown, whereas a strongly non-uniform field permits the development of a diffused steady discharge which contracts at much higher voltages when going over into a spark.

In the corona, ionization is confined to a very thin gas layer, in which electron multiplication by collisions due to the high field occurs. Such a system acts as a unipolar ion source provided that the electrons are quickly attached to molecules. Thus multiplication throughout the whole gas volume is impossible because of the rapidly decreasing field around the wire. Since the current flow is controlled by the ion mobilities in the low-field region, space charges accumulate there and prevent the field from rising swiftly near the wire. As a result of it a more gentle increase in current with rising voltage arises. The space charge effect also causes current oscillations, usually at larger voltages, and there is a tendency for coronas to develop space charge instabilities known as Trichel pulses.

The neutral dust particles in the gas flow encounter ions and acquire some of their charge. They are acted upon by the electric field which drives them towards the wall electrode where they come to rest and lose their charge. The neutralized particles are either collected by falling under gravity or are detached from the wall by mechanical means (vibrations). The radial motion (v_r) of the charged particles is controlled by the electric force (a force due to an induced or permanent electric dipole moment is negligible), by corona spot turbulences and by the viscosity of the gas. For small turbulence and a charge ve per particle of radius ρ, the radial velocity

$$v_r = veE_r/6\pi\rho\eta \tag{8.1}$$

where the viscosity of the gas of density δ is

$$\eta = \bar{v}\lambda mN_0/3 \tag{8.2}$$

where $mN_0 = \delta$ and the other variables have their usual meaning. For air at room temperature $\eta \sim 2 \times 10^{-4}$ poise (2×10^{-5} Pa s). With $v \sim 10^2$, $\rho \sim 1\,\mu m$, $E_r \sim 3$ kV cm^{-1} at $r \gg r_0$, we obtain $v_r \sim 5$ cm s^{-1}. The residence time in a tube of 5 cm radius for 1 m s^{-1} flow velocity is thus about 1 s.

For the particles floating in the gas, we shall distinguish between conducting and insulating ones. Assuming spherical shape, the latter will acquire a final charge ve when the field of a single particle equals that of the uniform applied field

$$E_r = \frac{ve}{\rho^2}\left[1 + 2\left(\frac{\varepsilon - 1}{\varepsilon + 2}\right)\right] \tag{8.3}$$

where ε is the dielectric constant. For conductors $\varepsilon = \infty$. It is convenient to divide particles into two groups: $\rho > 1\,\mu m$ for ash, cement and $\rho < 1\,\mu m$ for cigar-smoke, oil-smoke, for example. Figure 8.3 shows how the final value of v depends on ρ when $E_r = 2$ and 5 kV cm^{-1} and $\varepsilon = 2$. Equation 8.3 has been experimentally confirmed for $\rho > 1\,\mu m$. For small particles, theory and experiment give $v/\rho \sim 2 \times 10^5$ (dashed lines, figure 8.3). Adding dust particles to air reduces the corona current (figure 8.2) via the effective ion mobility.

The degree of precipitation $(1 - \xi)$ is related to the ratio of the number of dust particles per unit volume leaving the cylinder to the number entering it. Assuming radially uniform dust particle density N_p of the dust-laden gas which flows axially through a cylinder of radius R with speed v_z, the filter removes dN_p along $dz = v_z \, dt$ and hence per unit length

$$R^2 \pi \, dN_p = -2\pi R l v_r N_p \, dt \tag{8.4}$$

By integration over the filter length $l = v_z t$

$$\frac{(N_p)}{(N_p)_0} = \exp\left(-\frac{2v_r}{v_z} \frac{l}{R}\right) = 1 - \xi \tag{8.5}$$

In modern plants the fraction ξ leaving the top of the chimney is very small ($\xi \ll 1$). $(N_p)_0$ can be of the order of 10^8 cm^{-3}.

8.8. Gas discharge displays

One type of display panel, fitted with a large number of light-emitting letters, numerals or tubes forming a general pattern, consists of gas (glow) discharges. Each element is a miniature two-electrode cell, each representing a luminous point (or line), and they are arranged in dot or bar-matrix arrays (figure 8.4). The single cell of about 1 mm diameter or less is usually filled with a Ne–Ar (Penning) mixture at a pressure of the order 10 Torr. The cathode is a low atomic number metal or carbon to reduce sputtering, the anode is a metal and the inter-electrode distance is several cell diameters. The cell current is of the order $10 \, \mu$A. The starting and working voltages are about 250 and 150 V respectively and the luminance (brightness) a few thousand cd m^{-2} /1 cd (candela) ~ 1 international candle; 1 fL (foot-Lambert) $\sim 10^{-3}$ L; 1 cd m$^{-3} \sim \frac{1}{3}$ fL]. Because of the small size of the cell and the relatively large gas pressure the light is almost entirely emitted from the negative glow, and a positive column is absent. In addition to the d.c.-discharge cells discussed above, a.c.-operated cells arranged in panels have been developed with large plane external electrodes covering the cells' end-walls.

Figure 8.5 shows enlarged a panel element which consists of three glass plates each about 0·2 mm thick. The central plate 2 contains the gas-filled hole (cell) about 0·4 mm in diameter and a cell spacing of about 0·75 mm. On top of the central plate 2, with the array of holes rests plate 1, whose upper surface is covered with a transparent conducting layer (not shown). On the lower side of plate 2 is plate 3, which covers the holes; on its inner surface and thus in contact with the gas filling, sets of orthogonal transparent conductors are deposited

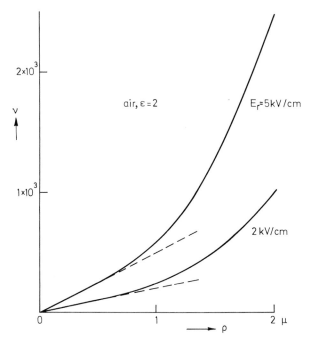

Figure 8.3. Final particle charge ν in units of e as a function of the mean particle radius ρ resulting from a corona in air containing insulating particles ($\varepsilon = 2$) for different fields E_r at the collecting cylinder.
Dashed lines: $\nu = f(\rho)$ for small particles. $e =$ electron charge.

(figure 8.6). In this way each cell can be separately operated. The plates 1, 2 and 3 are sealed together along the edges and the cells are filled with gas. The light emitted by a.c.-operated cells occurs in the form of pulses whose shape and frequency can be changed, for example by adding N_2 (<1 per cent) to Ne. At 50 Hz the duration of the light pulses is only a small fraction of a half cycle. Note that the plates 1 and 3 represent capacitances and hence no additional current-restricting circuit elements are needed.

The present development is directed towards an increase in luminance, number of cells per panel, life, gradual variation in luminosity and colour, and improvement in general operation. For switching circuits consult Further Reading.

Figure 8.4. 'Dot' and 'bar' matrix array.

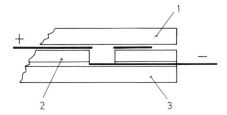

Figure 8.5. An a.c.-operated discharge cell with transparent conductors.

8.9. Flame radiation detection

A fire detector has been developed whose working principle is based on the detection of radiation from hot flame gases. By repeatedly measuring the statistical time lag of a gas discharge initiated between metal electrodes in a gas mixture of moderate pressure, it is possible to distinguish between background radiations, including the sun, and the intrinsic radiation from the hot gas.

It can be shown that at moderate gas pressures the total time lag τ of an incipient discharge is practically equal to the statistical lag τ_s because the formative lag $\tau_f \ll \tau_s$. The definition of the lag τ_s follows from observations of the probability that the electron released from the cathode will lead to breakdown of the gas mixture when an appropriate voltage V is applied to the gap. If in n_1 cases out of a total number of n_0 trials no breakdown occurs for a voltage pulse of duration t, it can be shown that

$$n_1/n_0 = \exp(-t/\tau_s)$$

where τ_s depends on the rate of photoelectric emission of the cathode and on $(V - V_b)$, the pulse height V in excess of the breakdown (or starting) potential V_b. The observed slope of $\ln(n_1/n_0) = f(t)$ gives τ_s. If the coefficient of secondary emission depends strongly on the photoelectric effect, i.e., little on the positive ions and metastables, then any radiation corresponding to $h\nu < e\phi$, ϕ being the work function of the cathode, is ineffective. This means that the quantum yield in photoelectrons per quantum absorbed is zero. Thus for nickel cathodes ($\phi = 4 \cdot 9$ V) only quanta of less than about 250 nm can release electrons.

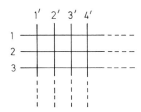

Figure 8.6. Cross bar array for starting selected cells. For example, one where the cathode bar 2 crosses the anode bar 3'.

Plane nickel cathodes and grid anodes have been used in mixtures of H_2 and Ar at pressures of 30–100 Torr and E/p of about 30 V cm^{-1} Torr ($E =$ electric field strength). Grid anodes are used to allow the radiation to fall with little hindrance on the cathode surface. τ_s is measured by integrating the current pulses resulting from electron avalanches in the gas which a constant voltage pulse produces. The repetition frequency of the pulses is <1 kHz and the number of avalanches is counted during a certain time interval t for a given intensity of u.v. radiation. Excess radiation indicating the presence of a flame can thus be detected rapidly in industrial plants and aircraft engines.

I shall not discuss cosmic radiation and particles and their measurement here, but some references on the subject are cited in Further Reading, notably the works of Breare, Sharpe, Stubbs and Breare, and Taylor.

CHAPTER 9

nuclear fusion plasmas

Plasmas in which two atomic nuclei are fused together are of interest in the quest for fusion reactors—the nuclear boilers of future power stations. The fusion reactions for deuterium ($_1^2 D$) and for deuterium–tritium ($_1^3 T$) read (figure 9.1)

$$_1^2 D + _1^2 D \rightarrow _2^3 He + _0^1 n + 3 \cdot 3 \text{ MeV} \tag{9.1a}$$

$$_1^2 D + _1^2 D \rightarrow _1^3 T + _1^1 H + 4 \text{ MeV} \tag{9.1b}$$

$$_1^2 D + _1^3 T \rightarrow _2^4 He + _0^1 n + 17 \cdot 6 \text{ MeV} \tag{9.2}$$

The ratio of the likelihood of the two reactions for D–D fusion is about $1:1$. Generally, in a fusion process energy is only released if the reaction products, such as $_2^3 He + _0^2 n$, have a smaller total mass than the primary one, here $D + D$. The reaction energy $\Delta\varepsilon$ is given by the absolute mass defect Δm (in g) with c, the velocity of light; the energy equivalent of the mass defect is

$$\Delta\varepsilon = c^2 \Delta m \tag{9.3}$$

Applied to equation 9.1a, $\Delta m / m_H \simeq [(2 \times 14 \cdot 7) - (17 + 9)] \times 10^{-3} \simeq 3 \cdot 4 \times 10^{-3}$ a.m.u.[†] Hence, with $m_H = 1 \cdot 67 \times 10^{-24}$ g, we obtain

$$\Delta\varepsilon = 3 \cdot 4 \times 10^{-3} \times 9 \times 10^{20} \times 1 \cdot 67 \times 10^{-24} = 5 \cdot 1 \times 10^{-6} \text{ erg} \simeq 3 \cdot 2 \times 10^6 \text{ eV}$$

in agreement with $\Delta\varepsilon$ of the first reaction equation above (1 a.m.u. $= 930$ MeV).

Fusion reactions release energy only when light nuclei join to form a heavier nucleus. This occurs for mass numbers below 60, i.e., elements lighter than nickel, but for practical reasons the choice of plasma substances is more restricted. The reaction energy appears as kinetic energy of the product nuclei which recoil in the inverse ratio of their masses. For example, in the D–T reaction $_2^4 He$ carries away 1/5 of the total energy, i.e., $3 \cdot 5$ MeV, while 80 per cent (14 MeV) go to the neutron, a result which follows from conservation of linear momentum. However, whereas 20 per cent of the energy in this case is associated with a charged particle,

[†] a.m.u. = atomic mass unit. 1 a.m.u. $\simeq$ mass of H $\simeq$ mass of neutron $_0^1 n$. 1 a.m.u. is defined as 1/12 of ^{12}C. For isotopic masses see Semat 1964; e.g., for $_1^2 D$, $m_D = 2 + 14 \cdot 7 \times 10^{-3}$ a.m.u.

175

Figure 9.1. Fusion of deuterium (1 p + 1 n) and tritium (1 p + 2 n) ions.
[DT]* = excited compound nucleus temporarily formed.

the remainder is with a neutron of high penetrability (small cross-section). To extract its energy a proton or lithium-rich 'blanket' could be used. Since natural Li contains 93 per cent $_3^7$Li and 7 per cent $_3^6$Li, a blanket bombarded by neutrons produces tritium

$$_3^6\text{Li} + _0^1\text{n} \rightarrow _2^4\text{He} + _1^3\text{T} + 4 \cdot 8 \text{ MeV}$$

$$_3^7\text{Li} + _0^1\text{n} \rightarrow _2^4\text{He} + _1^3\text{T} + _0^1\text{n} - 2 \cdot 5 \text{ MeV}$$

(9.4)

which can replace $_1^3$T 'burnt' in the D–T plasma at about the same rate as it is consumed in the blanket. $_1^3$T is a radioactive isotope with a half-life of about 13 years, decaying into $_2^3$He and e (18 keV maximum), whereas the half-life of $_0^1$n against β decay, i.e., $_0^1\text{n} \rightarrow _1^2\text{H} + \text{e} + \nu$ (neutrino), is about 12 minutes.

Let us now consider the physical conditions which are required for the D–T fusion reaction to occur at a sufficient rate, each individual fusion event giving $E \sim 18$ MeV energy. The power produced per unit volume for number densities N is

$$P = \varepsilon \overline{qv} N_D N_T$$

(9.5)

The rate constant $\overline{qv}$ depends sensitively on the plasma temperature T_p, often expressed in eV (1 eV $\sim$ 11 600 K), since very high relative velocities between D and T are necessary to bring the nuclei so closely together that fusion occurs. In fact, an excited intermediate $(_3^5\text{Li})^*$ nuclear state is formed during a collision; $(_3^5\text{Li})^*$ subsequently splits up into He + n. Figure 9.2 shows the power developed $P = f(T_p)$ in D–T when the number density $N = N_D = N_T = 10^{15}$ particles cm^{-3}. If the main power loss is by 'bremsstrahlung' P_{rad} only, i.e., if impurity and plasma convective losses are absent, the critical (or break-even) temperature $T_p = T_c$ is reached when $P = P_{rad}$ and, since $P_{rad} \propto T_p^{1/2} N^2$, we find from figure 9.2 that $T_c \sim 4$ keV $\sim 5 \times 10^7$ K. In practice the working value of T_p will be about 10^8 K to achieve satisfactory efficiency. In addition, an energy balance per unit volume of plasma leads to another relation which involves the product of the plasma density N and the least time, the confinement time τ, the plasma must be kept at $T_p \sim T_c$.

The parameter $N\tau$, which indicates whether a reactor will provide a net output of power, can be obtained by equating the energy required to produce unit volume of plasma at temperature T for τ seconds (energy confinement time), where N is the initial number density of particles, with the total energy output in

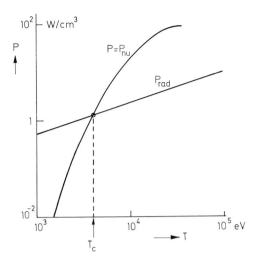

Figure 9.2. Power P per unit volume $= f(T)$.
$T =$ plasma temperature; $P_{rad} =$ radiation power loss; $T_c =$ critical temperature; $N_D = N_T = 10^{15}$ cm$^{-3} = 10^{21}$ m^{-3} ion concentration.

form of electric energy. If η is the efficiency in converting heat to electric energy which applies to the nuclear radiation (bremsstrahlung) and specific heat power terms P_{nu}, P_{rad} and $3NkT$ respectively, we obtain

$$P_{rad}\tau + 3NkT = \eta[P_{nu}\tau + P_{rad}\tau + 3NkT] \tag{9.6}$$

and thus, to deliver net power, Lawson's criterion must hold

$$N\tau > \frac{3N^2 kT}{[\eta/(1-\eta)](P_{nu} - P_{rad})} \tag{9.7}$$

Taking $\eta = 0.3$ and a D–T plasma, $N\tau > 10^{14}$ cm^{-3} s and $T_c \gtrsim 5$ keV, as above, because the denominator of equation 9.7 must be >0 to make $N\tau > 0$. At present, experimental values are within a factor of three of the required value of $N\tau$. Since both P_{nu} and P_{rad} are proportional to N^2, $N\tau$ depends on T. Figure 9.3 shows the observed ion temperature T; as a function of $N\tau$ obtained with various types of apparatus, indicating clearly that none of the methods used have up to now brought us across the theoretical limit given by the confinement of the plasma. Empirically in tokamaks (see p. 183) $N\tau$ values scale as the square of the minor radius of the torus (discharge vessel) which explains the huge size of the apparatus (e.g. JET, see p. 184).

So far we have only dealt with the collisions between, say, D and T, each of which will lose its electron at temperatures far below the critical, thus forming a quasi-neutral plasma. The question of how the charges consisting of D^+, T^+ and two electrons per D–T are balanced when fusion occurs into $^3_2$He $+ ^1_0$n is easily answered. The He isotope is a light slow α particle, namely He^{2+}. This point is of

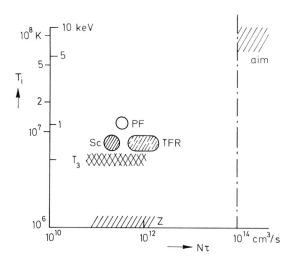

Figure 9.3. Ion temperature $T_i = f(N\tau)$, ion concentration × confinement time.
Z = Zeta, PF = plasma focus, T_3 and TFR = Tokamak, aim = net output > 0,
Sc = Scyllac.

some importance with respect to the containment problem. As we shall see below,
applied magnetic fields are usually used to keep the hot plasma away from the
walls of the vacuum vessel, but this can only affect the motion of charged
particles, whereas the neutron paths remain unchanged. The direction of flow of
energy carried by fast neutrons cannot be controlled.

Let us assume that the 'break-even region', indicated in the right corner of
figure 9.3, has been reached. There still remain at least two other main problems
to be solved, namely improving the confinement and avoiding instability. We shall
discuss first some of the methods used to keep the discharge, formed in the gas of
low density in which the fusion reaction (figure 9.1) takes place, from reaching the
walls of the discharge vessel. Otherwise, apart from the cooling of the plasma and
the damage which ablation inflicts upon the walls, the higher temperature
components of the plasma would release impurities which would greatly increase
the unavoidable losses by bremsstrahlung and reduce the life of the evacuated
container. A recent idea is to protect the inner wall surfaces with a covering layer.

There are only two practical solutions available, at present, to confining a
fusion plasma, namely either applying a magnetic field or laser compression,
sometimes called inertial confinement. In principle, the former is based on
Faraday's concept of the interaction between the magnetic field of the plasma
current and the externally applied field, attributing to the field lines the properties
of elastically stretched rubber filaments which, besides longitudinal contraction,
exhibit lateral pressure. In the simplest case of a uniform plasma with current
density j and field B in a bounded volume, the magnetic pressure gradient normal
to $\mathbf{j}$ and $\mathbf{B}$ is given by $\nabla p = \mathbf{j} \times \mathbf{B}$, the force per unit volume, and with $\mu_0 \mathbf{j} = \nabla \times \mathbf{B}$
we have $(1/\mu_0)\nabla p = (\nabla \times \mathbf{B}) \times \mathbf{B} = -\mathbf{B} \times \operatorname{curl} \mathbf{B} = -\frac{1}{2} \operatorname{grad} \mathbf{B}^2 \times \mathbf{B} \operatorname{grad} \mathbf{B}$. If $\mathbf{B}$ is

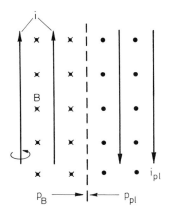

Figure 9.4. Magnetic confinement of a plasma.
Electron and ion plasma currents (with components perpendicular to the paper) cause the plasma pressure p_{pl} which is balanced in the boundary plane ($x=0$) by the lateral magnetic pressure p_B of the external field B at $x<0$. $\beta = p_{pl}/p_B$.

uniform, grad $\mathbf{B}=0$ and $\nabla p=0$. With the partial pressures $p_e + p^+ = 2p_e = p$ and $T_e = T^+ = T$, we find that in equilibrium, for complete ionization

$$p = \tfrac{1}{2}BH = B^2/2\mu_0 = 2NkT \qquad (9.8)$$

where $\mu_0 = 4\pi \times 10^{-7}$ H m$^{-1} \sim 1\cdot26 \times 10^{-6}$ H m^{-1} and $k = 1\cdot38 \times 10^{-23}$ J K^{-1}, i.e., the (total) plasma pressure is balanced by the magnetic pressure. For $N = 10^{20}$ particles m^{-3} and $T = 10^8$ K, we obtain from equation 9.8 $p \sim 3 \times 10^5$ J m^{-3} (or N m^{-2}) ~ 3 atm. This equilibrium, according to equation 9.8, demands a field $B \sim 9$ kG ($=0\cdot9$ T). Figure 9.4 illustrates this, assuming a plane vertical boundary between plasma and field.

In the case of laser compression, the general notion is as follows: when a plasma has given N_e and thus f_{pl}, the electron plasma frequency (see equation 5.3) is induced by electromagnetic radiation of frequency f from a laser, and the light is increasingly absorbed as f is raised, reaching a maximum when $f = f_{pl}$. Absorption occurs partly by 'inverse bremsstrahlung', i.e., free electrons in the plasma experience deflection (scattering) and net acceleration by interacting with bound atomic or ionic electrons. The coupling between the hot electrons and the ions produces ablation, and the mechanical reaction or implosion (compression) of the D_2 pellet. Since in dense plasmas $N_e = 10^{17}$ cm^{-3} and $f_{pl} \sim 3 \times 10^{12}$ Hz, strong absorption requires powerful infra-red lasers of $\lambda = c/f \sim 100\,\mu$m. In this way an already established and confined D plasma can be heated. However, both processes can be combined. Let the initial 'plasma' be a small solid particle (pellet) of D–T or D at very low temperature which is irradiated by laser pulses coming from all directons. The radiation pressure can compress the pellet to, say, ten thousand times the standard density, which is of order 10^{22} atoms cm^{-3}. Thus, if laser radiation of $\lambda > d$, d being the linear dimension or diameter of the spherical pellet, is absorbed by the electrons in the solid, the temperature rises to a value that will induce the nuclear reaction. The conversion of the solid into a plasma at high temperature and pressure should occur within a short time. By replacing the 'burned' DT pellet every 10^{-2} s the maintenance of a fusion plasma might be feasible. So far, encouraging calculations are available.

Next, some questions concerning plasma particle and energy losses of the envisaged fusion reactors should be briefly discussed. Since the size of a several thousand MW reactor is considerable, about 10 m and 3 m diameter of torus and plasma vessel, respectively, the interaction between the hot plasma and the wall is of importance. The (classical) diffusion given by electron–ion collisions is known to represent only part of the total diffusion loss across a magnetic field, because small-scale instabilities and plasma drifting radially outwards across B due to electric fields, caused by radial magnetic field gradients, are responsible for additional diffusion losses. However, it has been possible to reduce the latter considerably, for example by changing the design of the magnetic field structure. In spite of recent progress, the lost escaping plasma carries with it a huge amount of energy; special chambers (dumps) have to be provided to 'cool' down the portion of the plasma which does not contribute to fusion. However, by extracting the thermal energy from the dumped plasma the efficiency of the fusion reactor can be much increased.

Further, some interaction between the hot radiating plasma and the wall of the vacuum chamber is unavoidable. To reduce it to a reasonable level, 'magnetic diverters' are used to deflect a part of the plasma, which is rich in impurities, into external chambers from which cleaner plasma is returned to the main chamber. These impurities consist of evaporated or ablated constituents of the wall and of α particles. Besides purification, diverters can control the reactor output, act as pumps and cool the plasma, thereby locally reducing thermal stresses. It would be unwise to go here into details of the fusion reactor operation and design; drawings and data can be found in numerous publications, some of which are cited in Further Reading.

Finally, a few examples of conventional arrangements for trapping fusion plasmas by magnetic fields will be illustrated and discussed. Trapping confines the plasma to a part of the volume which is at a sufficiently large distance from the wall of the container; it prevents removal of wall material and the release of impurities, and reduces the losses by the unavoidable plasma leakage through for example, pumping lines, which are connected to the container. One method used to obtain high temperatures and radial confinement involves the 'pinch-effect', i.e., the radial contraction of a long 'fluid' cylindrical electric conductor by its own magnetic field. It was first observed by electrical engineers using large eddy currents for melting metals. This contraction is the result of uni-directional parallel 'current filaments' (or of their circular magnetic field lines surrounding the conductor forcing the filaments radially inwards, like stretched rubber rings). The critical current i_c, at which the cylindrical pinch starts, can be found from a relation similar to equation 9.8 by substituting $B = 2i\mu_0/r_0$ in equation 9.8, giving with $N' = 2\pi r_0^2 N$ (ions per metre length)

$$i_c \sim 2 \cdot 5 \times 10^{-8} (N'T)^{1/2} \tag{9.9}$$

with $r_0 = 2 \times 10^{-2}$ m, $N' = 10^{17}$ m^{-1}, $T = 10^8$ K, $i_c \sim 10^5$ A, independent of the current density distribution.

Eddy currents in a nearby conductor can produce a similar effect on a pulsed plasma, as can its self-magnetic field. Consider a current sheet in a plasma that is bounded by a metal wall, the direction of the current being parallel to the wall and its magnitude varying with time. The associated magnetic flux induces an electric field so that eddy (secondary) currents circulate in the metal. These penetrate it to a depth Δ according to $\Delta \propto (\sigma \omega \mu \mu_0)^{1/2}$, σ being the electric conductivity, ω the angular frequency and μ the permeability. For a Cu wall and a frequency $f = \omega/2\pi = 1$ MHz we obtain $\Delta \sim 7 \times 10^{-2}$ mm. It follows that the magnetic flux of the secondary current, which is approximately in a direction opposite to that of the plasma current (phase angle $\leq \pi$) gives rise to a force (or pressure) repelling the plasma current sheet from the wall; however, the confining force ceases when $di/dt \to 0$. The 'wall confinement' of a plasma has its electromagnetic analogue: a loose Cu ring through which a straight vertical iron core protrudes is accelerated upwards when the core is magnetized by a.c., or when a current pulse in a coil (below the ring) is fixed to the core.

It was shown in figure 4.10 that electrons in the magnetic field between two coils—a mirror field—can be 'contained' by axial reflection for relatively long periods. The reason why a plasma in such a field is subject to small particle losses is seen in figure 9.5, where the field structure outside the coils is given in more detail. Though the field at each coil's centre is high, a certain fraction of charges escape by virtue of their axial velocity component. However, considerable improvement is obtained when a magnetic multipole field is superimposed on the simple mirror field. Figure 9.6(*a*) shows a magnetic quadrupole field formed by two pairs of current-carrying wires (straight or slightly bulging), figure 9.6(*b*) shows the field lines in a plane B between the coils A and A', and figure 9.6(*c*) shows the resulting shape of the plasma formed. Instead of magnetic mirrors, a cylindrical plasma (again with a 'loss cone' at each end) can be produced and contained by a cylindrical current sheet (figure 9.7). The current pulse in the

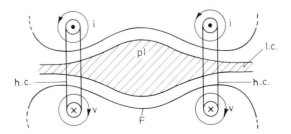

Figure 9.5. Mirror *B* field produced by a pair of Helmholtz coils (h.c.) illustrating the loss cones (l.c.). Coil currents in the same direction.

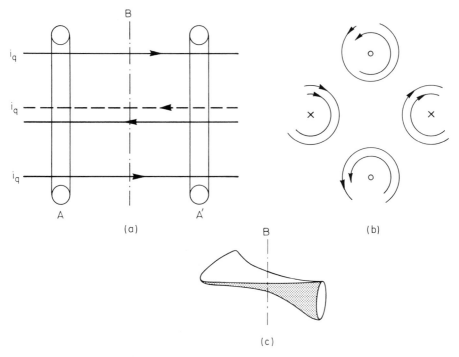

Figure 9.6. Joffe multipole field obtained by combining a mirror and a quadrupole field q.

    (*a*) Side-view of mirror coils and axial conductors.

    (*b*) q field in the plane B $(-\cdot-\cdot-)$.

    (*c*) Resulting plasma shape.

single turn surrounding the vessel produces 'secondary' azimuthal currents in the hot ionized gas—hence the name thetatron. Axial plasma losses can be avoided by connecting the vessel's ends, which lead either to a simple torus (doughnut), also called 'zeta' (zero energy toroidal assembly), or—when twisted into a figure of eight—to a 'stellarator'. The Tokamak is at present the favoured toroidal device

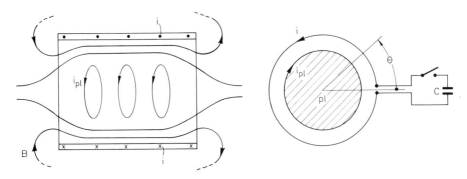

Figure 9.7. Thetatron, a cylindrical plasma contained by a cylindrical current sheet, pl = plasma.

(figure 9.8): a bakeable metallic vacuum chamber surrounds a central iron core which is pulse-magnetized by energizing the primary 'transformer windings'. The plasma ring discharge in the torus gas ($p \lesssim 10^{-2}$ Torr of D–T) represents the secondary winding. Several auxiliary magnetic fields are applied: a stabilizing longitudinal field B_z is set up by a coil wound around the whole circumference of the torus which, together with the self-field B_θ of the plasma current, gives a closed helical field (figure 9.8). The stabilizing field B_z used is of the order 3–5 T. Since plasma currents i_{pl} are about 0.1–5×10^6 A peak, B_θ max is of about the same order as B_z or smaller. In this way various plasma instabilities which are accompanied by unwanted motions and distortions of the plasma column have been successfully eliminated. It would be unwise to discuss details or modes of operations except to point out that, with the auxiliary field, energy containment

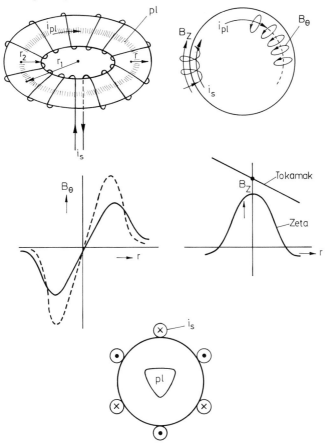

Figure 9.8. Tokamak: evacuated torus (from the Russian acronym to-ka-mag = toroidal-chamber-magnetic).
This contains the plasma, toroidal (axial) component B_z caused by coil currents, and poloidal (azimuthal) component B_θ due to the net plasma current i_{pl}. $B_z > B_\theta$. Bottom: cross-section of torus, top left.

times $\tau \sim 0.15$ s have been achieved. Particle concentrations of up to 10^{14} cm^{-3} and kT_i and kT_e of about 6×10^3 eV($\sim 7 \times 10^7$ K) have been observed, indicating that considerable advances have been made.

The European tokamak JET (Joint European Torus) will be used soon (1983) to study fusion reactions in D–T plasmas heated by a plasma current supplemented by radio-frequency heating and heating by injecting beams of neutral hydrogen or deuterium atoms of high energy (H$^+$, D$^+$ neutralized by electrons).

In addition, the engineering and economic aspects of fusion and fission–fusion reactors will be studied, in particular the problem of efficiently moderating the fast neutrons that are emitted by the D–T plasma by providing a lithium compound blanket surrounding it. The corresponding heat flow is about 80 per cent of the nuclear energy released. The heated blanket can be linked with heat exchangers supplying the heat flow to gas or steam turbines which produce electric power by conventional means.

chemical changes in discharges and plasmas

10.1. General considerations

Electric discharges have been known for over 400 years to cause chemical changes in gases or gas mixtures. These changes can consist either of the breaking of chemical bonds, such as the dissociation of diatomic molecules into their constituent atoms, or the rearrangements of bonds, leading to the formation of compounds from elements or molecules. The chemical effects of a lightning stroke or of sparks in air as produced by a Wimshurst machine are examples. One of the earliest quantitative experiments, conducted around 1910, which we shall discuss, is the synthesis of water from hydrogen and oxygen in an ionized gas at room temperature; note that the first electrostatic machines were used in the 17th century.

Today, in large-scale plants, effluents and wastes of textile, animal and vegetable matter are oxidized. Extensive installations exist for producing ozone (O_3), particularly for the purification of water; yet the power required is relatively small. In contrast, long, air-blown, high-current arcs between cooled Cu electrodes, disc-shaped by a strong magnetic field, produce about one per cent NO in air. When the cooled gas is oxidized to NO_2 and passed over water, nitric acid (HNO_3) forms; this (Birkeland–Eyde) process is not used any more. Similarly, high-power arcs between carbon electrodes acting on CaO (quicklime) give calcium carbide CaC_2. In the latter cases, the high gas temperature in the discharge column is essential to obtain the reactions.

In general, the ionized gas is mostly in the form of a long cylindrical column through which an electric current passes. Assume that the gas pressure and the current are so low (1 Torr, 10 mA, tube radius 2 cm) that no appreciable heat is produced in the gas. Since the number of electrons per unit volume is very nearly equal to that of the positive ions, the electric current which traverses the cold gas is mainly carried by the highly mobile electrons, whereas the role of the slow, heavy, positive ions, which usually carry less than one per cent of the total current, is essentially to annul the electron space charge.

At first sight such a system seems to be closely analogous to a liquid

185

electrolyte, in which positive and negative ions move. Yet in the majority of gases electrons are not transferred between atoms or molecules, and negative ions are therefore absent. It is therefore not surprising that the science of electric discharges in gases, which emerged from the study of electrolytic processes, initially used the electrochemical nomenclature. However, words such as 'cations' (+) and 'anions' (−) are no longer in vogue in the field of gas ions.

Because of the apparent analogy between the electric conduction in electrolytes and in gases the first question which springs to mind is whether Faraday's law holds for chemical reactions in ionized gases. Faraday's law of electrolysis states that the mass of gas liberated at the electrodes, or the mass deposited on the cathode, is proportional to the product of current and time, i.e., the electric charge which has passed through an electrolyte. The observation that the mass of the reaction products of an electrochemical process in a gas is proportional to the charge passed is not sufficient to confirm Faraday's law. For it can be shown (*a*) that other parameters, such as the electric field and the gas density, have a controlling influence on ionized gas reactions, and (*b*) that the mass of the reaction products is not determined by Faraday's constant. According to Faraday's law, 1 g-equivalent of any ion, say 2 g of H_2 or 32 g of Cu (Cu^{2+}), corresponds to a transferred charge of 96 540 C. The physical reason why this law cannot be applied to the 'electrochemistry of gases' is that in ordinary liquid electrolytes 'spontaneous dissociation' into positive and negative ions takes place while the electrolyte is prepared (e.g., a salt dissolved). The energy for separating the ions is abstracted from the surrounding liquid in the form of thermal energy. Spontaneous dissociation is thus caused by the frequent collisions between solvent (liquid's molecules) and solid particles, and this depends on the force at impact. Its magnitude will be lower the larger the dipole moment (dielectric constant) of the solvent. The sufficiently energetic molecules form a temporary bond on impact, thereby removing from the crystal surface atomic ions which enter the electrolyte. An external source—a battery—provides the voltage necessary to move the ions against the frictional force which they encounter by traversing the liquid,

The gas, however, is originally an insulator which becomes a conductor during the starting period of a gas discharge. To maintain this state, energy has to be supplied not only to move the charges through the gas but to separate electrons from some of the neutral molecules, i.e., to balance the level of ionization against charge losses. Of particular interest is to compare Warburg's (1904) measured discharge yields of ozone formation for flowing O_2 with that by electrolysis. It was observed that at 300 K only about 500 C per O_2 gram equivalent (32 g) are needed in a 'silent' (corona) discharge as compared with 96 500 C per gram equivalent according to Faraday's law.

Another example is the chemical equilibrium of the system

$$2CO + O_2 \rightleftharpoons 2CO_2$$

The equilibrium concentrations of the three components are found to be the same whether a gas temperature of 2600 K or a corona discharge at 300 K is active. The physical reason for this remarkable observation will be given later.

The second question, equally fundamental in nature, is concerned with the applicability of the law of mass action: this relates the concentrations of the reactants and products to a constant which depends on the temperature and the activation energy (or potential well) of the reaction. Again, it has been found that the electrochemical changes in gases at room temperature do not obey the law of mass action or classical thermodynamics. We therefore conclude that intrinsically different physical and and chemical processes must take place in the gas of electric discharges. Ultimately, the difference is that electrolysis and thermochemical equilibria are related to thermodynamics, whereas discharge chemistry at moderate temperatures is related to rate processes, modified by reverse reactions and by vibrationally and electronically excited species.

What is the nature of these processes? The short answer is that in the large majority of cases, unlike in electrolytes, free electrons and singly charged positive (and sometimes negative) ions are formed in the gas by electron collisions. The applied electric field in the gas, which is very much stronger than that in electrolytes, drives the electrons in the direction of the anode over 'free' distances larger than those in electrolytes while the heavy positive ions move slowly to the cathode. When electrons collide with neutral particles they are scattered in all directions but lose little energy. Thus some of them acquire in the field large velocities and energies. A certain number have energies sufficiently large to excite and ionize gas molecules. The excited particles can either emit light or combine chemically with other particles forming compounds and radicals, other excited particles may disintegrate into neutral atoms, i.e., dissociate. Though these final products are often uncharged, some may carry with them, for a short time, potential energy in the form of vibrational or electronic excitation; we shall not discuss the small rotational energy because it is readily converted into kinetic energy. As a result of the numerous collisions, the electrons between them form an ensemble which is called an electron gas, a concept which has to be discussed in more detail.

Electrons are light small particles. About 2000 to 200 000 electron masses are equal to one atomic mass. The average velocity of electrons $\bar{v}_e$ in thermal equilibrium with the gas, i.e., in zero electric field, is therefore larger than that of gas molecules ($\bar{v}$) by the square root of the mass ratio; that is $\bar{v}_e$ is at least 50 to 500 times larger than $\bar{v}$ and the electron collision rate is much larger too. In the presence of an electric field the electrons are accelerated and suffer mainly elastic collisions so that the electron gas is 'overheated'. In the steady state, it can acquire temperatures (or mean energies) which may be a hundred times or more that of the neutral gas or the ions. Such conditions prevail, for example, in cold fluorescent lamps where the gas and the glass wall are not much above room temperature of 300 K while the electron temperatures often rise up to 30 000 K and more, or one hundred times room temperature. This argument holds only when both electron and gas densities are sufficiently low.

How can we explain that two so-called gases, which are thoroughly mixed, can be kept at such widely different temperatures? The answer is that in a steady state, as distinct from thermodynamic equilibrium, energy flows from an energy source through the electric field to the electron gas (neutrals cannot draw energy from an electric field and ions only little). As a result of the collisions, an energy distribution of electrons is formed, characterized by a large mean energy compared with that of the molecules, say 3 eV compared with 0·03 eV. The majority of the electrons colliding with the gas molecules, because of their relatively small mass, can transfer to the molecules only a minute fraction of their energy in a single collision.

The total energy which the neutral gas will receive is proportional to the electron concentration and the collision frequency, the latter being proportional to the gas density. Therefore at low gas pressure, as in many discharge lamps, the heat transferred from the electrons to the gas is small and the glass envelope remains cold, whereas at high pressure, as in open welding arcs or in high-pressure sodium and mercury lamps, the gas will be heated up to many thousand K and the inner envelope somewhat less. The reason why electrons can be overheated to a large extent is simply that at low pressure their thermal contact with the neutral gas is very poor indeed. Though mainly work in the low range of pressure (0·1 to 10 Torr) will be discussed here, I shall indicate the conditions which would arise if the pressure were increased to one atmosphere. We may think in some respects of the 'electron gas' as if its properties were that of an ordinary neutral gas. It will be necessary to find the number of electrons that have sufficient energy to excite neutral molecules which execute chemical reactions. We have to take into account that the 'activity' (cross-section) of electrons with larger energies may be greater than of those with lower energy.

In early fundamental work (see Chapters 4 and 7) photo-electrons were released from a metal cathode by shining ultraviolet light on it. These electrons move through the gas towards the anode when a driving force is provided by an electric field E. Some of the electrons collide frequently with the neutral gas molecules and are scattered at random but, at the same time, they are accelerated in the field direction. This led to the concept of a drifting 'electron swarm'; its motion resembles that of a swarm of bees, characterized by a large average random velocity plus a small average drift velocity (figure 4.24). However, in larger fields, charge reproduction, i.e., multiplication of electrons (and positive ions) takes place because of the presence of electrons with energies above that required for ionization. This is shown in figure 7.6 (p. 129) where the final electron energy distribution dN/N is plotted over the energy ε; yet only electrons in the shaded 'tail' of the curve are effective in producing new secondary charges in pairs. It is convenient to define a coefficient of ionization α_i which gives the number of newly created electrons and positive ions which a single electron produces when it moves one centimetre in the E (field) direction. Since for single collisions, $\alpha \propto N$, the gas density, the coefficient per Torr pressure, α/p, is often used at constant temperature. α/p is a function of the mean energy $\bar{\varepsilon} = eE\lambda_e$

which an electron has acquired along one mean free electron path λ_e along E. For a given temperature, $\lambda_e \propto 1/p$ and hence the mean energy $eE\lambda_e \propto E/p$. These are Townsend's concepts, developed soon after 1900, that are still the basis of the multiplication of charges by electron collisions in gases.

10.2. The synthesis of water from H₂ and O₂

One of Townsend's collaborators, Kirkby, thought that similar ideas might be applicable to chemical changes. The problem he studied was the synthesis of water

$$2H_2 + O_2 \rightarrow 2H_2O$$

Instead of using an apparatus similar to that of Townsend (see Chapter 7) with currents of the order 10^{-9} A, Kirkby used a glow discharge with current of up to 1 mA. He measured the time rate of change of pressure dp/dt in a closed cylindrical volume ($R = 1\cdot2$ cm) filled with two moles of H_2 and one mole of O_2 at pressures between $0\cdot2$ and 2 Torr (figure 10.1). By enclosing the discharge in a long tube the reaction occurred mainly in a uniform positive column. In order to exclude electrode and other end effects from the reaction, he measured dp/dt for different distances between the two electrodes and avoided any reverse reactions by restricting the time t to less than 12 s. The results of his work were most impressive. He found it convenient to define an association coefficient (α_{ass}/p) of

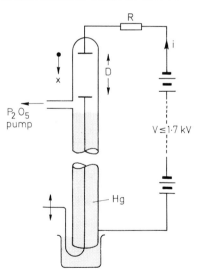

Figure 10.1. H₂O synthesis in a discharge column (Kirkby 1911).
D = electrode separation, varied by moving vertical U-shaped stem in the Hg-filled tube and beaker, thus forming a vacuum seal.

water that was absorbed in a side arm by P_2O_5; (α_{ass}/p) was shown to depend on E/p only, independent of current, time, etc., as was α/p (see figure 10.1). For example, in a field $E = 20$ V cm^{-1} at 1 Torr, $\alpha_{ass}/p = 1$, i.e., one water molecule per electron per centimetre of path per Torr in the E direction is formed. Note that the value of α_{ass}/p was (50 years later) found to be of the same magnitude as the dissociation coefficient, α_{ass}/p, of hydrogen, both at $E/p = 20$ V cm^{-1} Torr^{-1}. Kirkby concluded (a) that the laws of mass action and of Faraday do not apply, (b) that the primary agents which control the chemical change are electrons and not positive ions, and (c) that uncharged particles are chemically involved in the mechanism leading to the synthesis of water. As for ionization, the electrons are the chemically active species here, whereas the positive ions do not play any part in the reaction.

As we now know, Kirkby was, in a way, lucky in confining himself to this part of the problem, since the synthesis of water proved to be a fairly complex process. The present view is that first (swarm) electrons of energy not less than $8 \cdot 8$ eV dissociate H_2 molecules, and then atomic hydrogen and molecular oxygen form the radical HO_2 (detected mass-spectroscopically) but not OH directly. Thus at room temperature the following reactions take place

$$e + H_2 \rightarrow H + H + e \tag{10.1}$$

$$H + O_2 \rightarrow HO_2 \tag{10.2}$$

$$HO_2 + H_2 \rightarrow H_2O + OH \tag{10.3}$$

and also

$$H + OH \rightarrow H_2O \tag{10.4}$$

i.e., water is formed. Only a negligible amount of H_2O_2 is obtained in equilibrium. However, details of these processes, such as the vibrational state of O_2 leading to dissociation into $O + O^*$, are still missing. Yet the dissociation of O_2 by electron collision seems here to be quantitatively negligible.

This view, which is based on the presence of HO_2, is supported by a comparison of the dissociation rate coefficient α_{diss}/p as a function of E/p with the synthesis rate coefficient $\alpha_{ass}/p = f(E/p)$, as well as mass spectroscopy. As is seen in figure 10.2 the curves of the two coefficients α_{diss}/p and α_{ass}/p, plotted as a function of E/p, practically coincide. (For comparison, α_i, the ionization coefficient, is also shown; clearly $\alpha_i \ll \alpha_{diss}$.) This means that the dissociation of H_2 (but not O_2) is the rate-determining process in the synthesis of H_2O from gaseous H_2 and O_2 in the positive column of a glow discharge at low pressure. This is, of course, only true if reverse processes, such as dissociation of H_2O in a discharge, can be neglected. This can be achieved either by restricting the time of the electrochemical action or by passing a sufficiently fast flow of gas through the discharge region so that the 'residence time' t of the gas mixture in the column is small enough. In Kirkby's work, with the data given in figure 10.2, t is of the order 10 s. It also follows that the apparently 'obvious' steps in the synthesis of

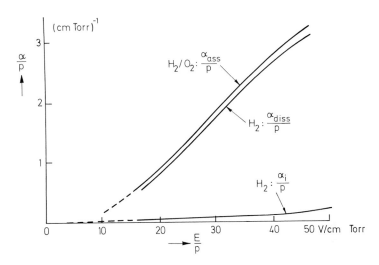

Figure 10.2. Synthesis of H_2O. Association and dissociation rates in a Townsend discharge. (From data by P. J. Kirkby (1911) and S. J. B. Corrigan and A. von Engel (1958) respectively; see Further Reading.)
α_{ass} from $1/i\,\Delta x\,dp/dt$ with $i \simeq 10^{-4}$ A, $t = 5$–12 s, $p = 0.2$–2 Torr, $D \simeq 30$ cm, $R \leq 3$ MΩ, $\alpha_i/p = $ ionization coefficient in H_2.

H_2O, namely $O_2 + e \rightarrow O + O + e$ followed by $H_2 + O \rightarrow H_2O$ do not take place. Instead, the unexpected radical HO_2 is formed, which interacts quickly with H_2 and produces H_2O. This is consistent with the large electron cross-section q_{diss} of H_2 relative to O_2 (compare figures 10.4 and 10.6).

10.3. Synthesis of ozone and dissociation of oxygen

Whereas the synthesis of H_2O is of purely academic interest, that of O_3 also has practical applications, especially for sterilizing fluids, since it is known that bacteria are destroyed by relatively small amounts of highly oxidizing O_3. The Earth is surrounded by an ozone layer which has a maximum density of ten times that at sea level at a height of about 20 km above the surface. It strongly absorbs the Sun's ultraviolet radiation in the range $\lambda = 200$–300 nm and acts as an optical filter.

The O_3 molecule is usually synthesized by a weak electric discharge of relatively short length. Electrons colliding with O_2 split some into O atoms which partly associate with O_2. However, whereas the binding energy of O_2 is 5·1 eV, the energy binding the third atom should be smaller than 2·5 eV and is, in fact, 1 eV. To achieve good efficiency, any excess energy has to be partially or fully taken away by a third body, such as O_2, within a vibrational period of time $\tau \simeq 10^{-12}$–10^{-13} s. Since a molecule collides with others $\bar{v}/\lambda$ times per second, one expects at room temperature $4 \times 10^4/6 \times 10^{-3} \simeq 10^7$ collisions per second at 1 Torr but about 10^{10} per second at 1 atmosphere. In the latter case not more than one per cent of the collisions will stabilize an O_3 molecule, which may be one of the reasons why the equilibrium concentration of O_3 is low, especially at low pressure.

The two steps which lead to O_3 when O_2 flows through a positive column of a glow (or silent) discharge are thought to be analogous to the synthesis of H_2O: (a) dissociation of O_2 and (b) association of O and O_2. Hence either

$$O_2 + e \rightarrow O + O + e \tag{10.5}$$

or

$$O_2 + e \rightarrow O + O^* + e \tag{10.6}$$

This is followed by

$$O + O_2 + (O_2) \rightarrow O_3 + (O_2) \tag{10.7}$$

where (O_2) indicates the 'third body'. It is clear that when the discharge extends only over a short length of tube, O_3 is formed mainly outside the discharge downstream, as has been observed at pressures of the order 1 Torr. Dissociation into O^* by reaction 10.6 seems to be more likely than by reaction 10.5 because, though a larger excess energy has to be removed (3 eV instead of 1 eV) in the former case, the Franck–Condon range favours reaction 10.6. This problem, however, is not yet solved. Another pair of possible electron collision processes are

$$O_2 + e \rightarrow O_2^* + e \tag{10.8}$$

and, excluding wall (heterogeneous) reactions

$$O_2 + e \rightarrow O_2^v + e \tag{10.9}$$

giving finally

$$O_2^v + O_2^* + (O_2) \rightarrow (O_4^*) + (O_2) \rightarrow O_3 + O_3 \tag{10.10}$$

(O_2) being again the third body, O_2^* an electronically excited metastable, O_2^v a vibrational state and (O_4^*) an intermediate state of low stability. (Note that O_4^+ ions have been detected mass-spectroscopically.) The reaction 10.7 involving (O_2) is thus associated with a reaction rate proportional to p^2. It is also known that the equilibrium concentration of O_3 decreases as the temperature is raised. O_3 can be frozen out below 160 K. In equilibrium we have an overall reaction

$$3O_2 \rightarrow 2O_3$$

The reason for (O_3) being sensitively dependent on temperature is not likely to be the low binding energy of O in O_3 of 1 eV consistent with the structure of O_3: the three atoms form a triangle (equilateral for reasons of symmetry) with an interatomic distance larger than that of O_2. The temperature dependence of O_3

formation may be due to collisions with $O_2^*(\Delta_g)$ which has a potential energy of about 1 eV.

Technically, the so-called silent a.c. discharge is mostly used in 'ozonizers' which add, to a flow of air or O_2, up to 10 per cent O_3 at 1 atmosphere. Figure 10.3 shows an ozonizer consisting of a concentric cylindrical gap through which, for example, O_2 is passed while a 50 Hz 'electrodeless' discharge is maintained. This is done by applying 5–20 kV between the large external electrodes of metal (tin) foil, which are attached to the glass so that a fairly uniform corona discharge (avoiding sparks which would decompose O_3) develops. Less than 15 per cent of the electric power supplied is actually required for the chemical change; the main part produces gas heating. The O_3 yield in O_2 is said to be about 4×10^{-5} g J^{-1} provided the flow is fast and the gas temperature kept low, as otherwise reverse reactions set in. The critical reader who compares the energy yield with the value of about 16 C g^{-1} given in Section 10.1 may conclude that about 2 kV are required for this process. Since the applied a.c. voltage is of that order, differences must be attributed to the small voltage drop in the glass wall, the inaccuracy of earlier measurements and the fact that the in-phase (ohmic) current component is not sinusoidal but is a sequence of pulses.

The dissociation of O_2 into neutral atoms by electron collisions which can occur via the $A(^3\Sigma_u^+)$ state (reaction 10.5) is associated with a critical energy of about 5·5 eV resulting in two ^{3}P atoms; or it can take place via the $B(^3\Sigma_u^-)$ state (reaction 10.6) with about 7.5 eV onset energy which gives one ^{3}P plus one ^{1}D atom. The energy dependence of the cross-sections q_A, q_B, which have been determined only approximately, are shown in figure 10.4. The graph includes the sum of the vibrational cross-sections q_{vib} of the O_2 ground state $(^3\Sigma_u^-)$, q_a the cross-section of the metastable (Δ_g) state, and $q_{diss_{att}}$ the ion $O^-(^2P)$ plus an $O(^3P)$ atom, provided the energy is 19 eV (for other dissociation mechanisms see Capitelli, ch. 3).

It must be emphasized that the excitation of the ground-state configuration of O_2 is still a matter of some dispute because of the strong interaction between the

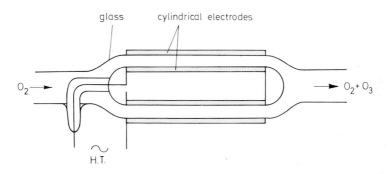

Figure 10.3. An ozonizer.
'Electrodeless' discharge in O_2 or air in glass tube with external electrodes by applying H.T. a.c.

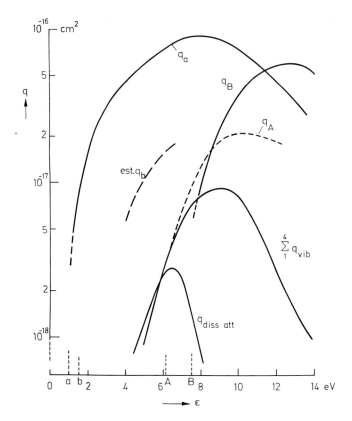

Figure 10.4. Various cross-sections $q = f(\varepsilon)$, and their dependences on the electron energy of O_2. q_A and q_B = dissociation via the A and B state respectively; q_a and q_b = excitation to $^1\Delta_g$ and $^1\Sigma_g^+$ metastable states respectively (see figure 10.5).

electronic ground state ($^3\Sigma_u^-$) and the O_2^- ($^2\Pi$) state (figure 10.5). We note the intersecting potential energy curves and the associated resonances that have been recently studied (see Further Reading). As a result of the coupling the representation of the transition, the path (vertical?) taken and the role and fate of the temporarily attached electron require more clarification, and the same applies to the quantitative side of the observed spectra, of the dissociation processes and of the electronic states of the products.

The electronic excitation of O_2 followed by dissociation can also be presented in a different way. Instead of considering only vertical transitions within the Franck–Condon region one could argue that any interaction between the $O_2(X^3\Sigma_g^-)$ and $O_2^-(X^2\Pi_g)$ systems during excitation to higher electronic states of O_2 is associated with a temporary displacement of the line 2 (figure 10.5) towards larger interatomic distances. In this case the energy-dependent transition path, starting at $X(v=0)$ would be expected to swing first to the right followed by a nearly vertical rise resulting from the weakening interaction.

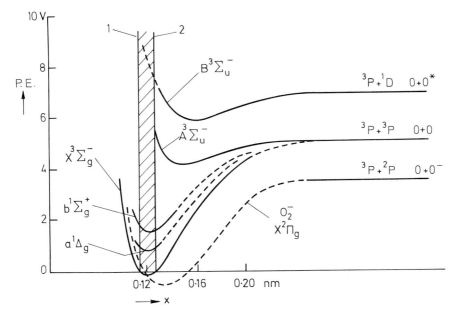

Figure 10.5. Potential energy $P.E. = f(x)$, the atomic separation x of O_2 and O_2^-.

The position of the O_2^- system relative to O_2 is consistent with the view that the attachment of an electron to O_2 must be accompanied by an 'opening' of its structure which agrees with the potential energy curves shown.

10.4. Dissociation of hydrogen

This is probably one of the best understood reactions. The potential energy curves of hydrogen are shown in figure 2.11 (p. 20). Figure 10.6 shows that an electron of $\varepsilon \geq 8\cdot8$ eV hitting an $H_2(A^1\Sigma_g)$ ground-state molecule and exciting it to the lowest repulsive electronic state $(b^3\Sigma_u^+)$, or to $(a^3\Sigma_g^+)$ decaying into the b state, can cause it to dissociate into $H + H$. Each atom then acquires a kinetic energy of about 2 eV since the thermal dissociation energy $D = 4\cdot5$ eV and the electron energy at onset is 8·8 eV. The cross-section q_{diss} varies with ε as shown in figure 10.6. Since this excitation is a singlet–triplet transition requiring an electron exchange, q_{diss} is expected to reach a maximum at $\varepsilon \sim 1\cdot5$–2 times the onset energy—which is seen to be the case. Since the lowest level of H^* is about 10 eV, excited atoms are ruled out as dissociation products at low ε.

When electrons move in an electric field through H_2 an electron swarm is formed and the corresponding dissociation can be described by a coefficient α_{diss}/p (in molecules dissociated per centimetre per electron per Torr) which has

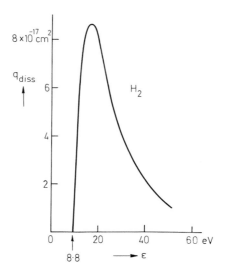

Figure 10.6. Electron collision cross-section $q_{diss} = f(\varepsilon)$ for H_2.

been measured (figure 10.7). It is related to q_{diss} (figure 3.14, p. 48) by R, the rate integral for dissociation, in events per second per electron

$$R = \int_0^\infty \left(\frac{q_{diss}}{q_{tot}}\right)\left(\frac{v_e}{\lambda_e}\right) f(\varepsilon)\, d\varepsilon = \alpha_{diss} v_d = \overline{q_{diss} v_e} N_0 \qquad (10.11)$$

where the first bracket is the dissociation probability, the second the collision frequency, and the product of the first and third terms is the relative number of 'active' electrons in the distribution, whose type must be known. For a

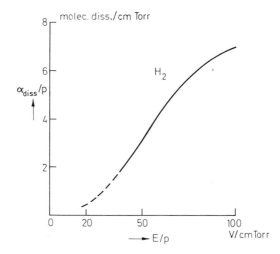

Figure 10.7. Dissociation coefficient $\alpha_{diss}/p = f(E/p)$ for H_2 at larger fields E/p.

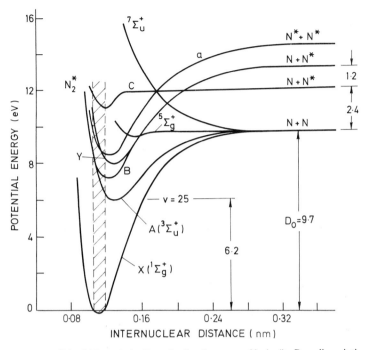

Figure 10.8. $P.E. = f(x)$ of N_2 and the Franck–Condon range (dashed); D_0 = dissociation energy in eV.

Maxwellian, $f(\varepsilon) \propto (\varepsilon/\bar{\varepsilon})^{1/2} \exp(-\varepsilon/\bar{\varepsilon})$ and with $q_{tot} \lambda_e = 1/N_0$, we obtain $R \simeq N_0 \int_0^\infty v_e q_{diss} f(\varepsilon)\, d\varepsilon$. (The lower limit is effectively $\varepsilon_0 = 8\cdot 8$ eV, since $q_{diss} = 0$ for $0 \le \varepsilon \le \varepsilon_0$.) For $E/p = 100$ we find $\bar{\varepsilon} \sim 5$ eV (see figure 7.7, p. 130) and from figure 4.24 (p. 77) $v_d = f(E/p)$. Thus equation 10.11 gives $R \simeq 8 \times 10^6$

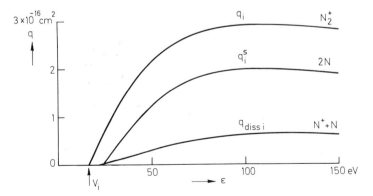

Figure 10.9. Ionization and dissociation into N^+; cross-sections $= f(\varepsilon)$.
q_i^s = super-excited state involved, see equation 10.12; $q_{diss\,i}$ = dissociative ionization cross-section.

(s electron)$^{-1}$, $(\alpha_{\text{diss}}/p) \simeq 3$ and $\overline{q_{\text{diss}} v_e} \simeq 2 \times 10^{-10}$ cm^3 s^{-1}, values which agree satisfactorily with observations.

10.5. Dissociation of nitrogen

The fairly complex electronic structure of N_2 which is represented by its numerous frequently intersecting potential energy curves (figure 10.8) and the absence of a repulsive state at low energy, indicates that the situation here is very different from that we have encountered with the relatively simple H_2 molecule. It follows that electron collisions are likely to give a large variety of electronically and vibrationally excited N_2 molecules which do not directly lead to dissociation. A positive column in N_2 at pR of the order 1 Torr cm^{-1} is known to emit the first and second positive bands showing that the $C(^3\Pi_u)$ and $B(^3\Pi_g)$ states are well populated; these decay finally into the metastable $^3\Sigma_u^+$ state. Under usual conditions, N_2^+ ions prevail but at lower pressure, i.e., larger $\bar{\varepsilon}$ and E/p, N^+ ions should and do appear whereas at higher pressure N_3^+ and N_4^+ have been found.

Direct single-step dissociation of N_2 into two $N(^4S)$ atoms is rare because swarm electrons of $\varepsilon > 24$ eV are scant. Ionization processes, such as dissociative and excited state ionization, may also yield N.

$$e + N_2 \rightarrow N^+ + N + 2e \ldots (q_{\text{diss}\,i})$$

$$N_2^{*s} + e \begin{cases} N^+ + N + 2e) \ldots q_i^s \\ \Big\}(q_i^s) \\ N + N + h\nu \end{cases} \text{ or } N^* + N + h\nu \tag{10.12}$$

where N_2^{*s} means a 'super-excited state' with a life of the order 10^{-5} s and an

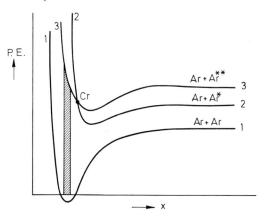

Figure 10.10. Dissociative excitation of a diatomic molecule (schematic).
Cr = crossing point; shaded area = Franck–Condon range; 1, 2, 3 = ground and higher states of electronic excitation.

energy of a few volts above eV_i in the continuum. N_2^{*s} has two decay routes. At present it appears that the second is the principal path, with neutral unexcited atoms as the main dissociation products. The energy dependences of the various cross-sections q_i, q_i^s and $q_{\text{diss }i}$ are shown in figure 10.9 but the values of q_i^s are still somewhat uncertain. One reason is the contribution by pre-dissociation processes, which are caused by the crossing of two potential energy curves (figure 10.10). This enables an excited molecule to change at the crossing point 'Cr' at constant energy from one state to another, the second having a lower dissociation energy than the first state. The result is a dissociation into atoms with kinetic energy: if one of the atoms is in an excited state the process is termed 'dissociative excitation'.

10.6 Dissociation of carbon dioxide

The electronic ground state of the triatomic molecule CO_2 has a linear structure with C at the centre (see Chapter 11). Each O atom has a binding energy of about 1 eV. The potential energy of the various electronic states of molecules with more than two atoms can only be represented by 'energy surfaces' and not by two-dimensional potential energy diagrams. There are only a few polyatomic molecules such as CO_2 whose energy surfaces are known (figure 10.11).

When a relatively slow electron collides with CO_2, the linear structure suggests a dissociation reaction of the form

$$e + CO_2 \rightarrow CO + O + e \tag{10.13}$$

However, originally this was not studied (at the time there was an obsession with ions, mass spectrographs, and imagined difficulties in detection of neutrals).

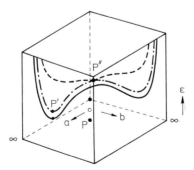

Figure 10.11. Potential energy surfaces of CO_2 and CO_2^* in a, b and z direction ($z \perp a$, b).
The curves on the front face representing CO show that the ground slate surface of CO_2 lies above that of CO_2^*. Thus a ball released at P' will run down towards P, overshooting it slightly because of the steep rising part of the surface near the origin O.

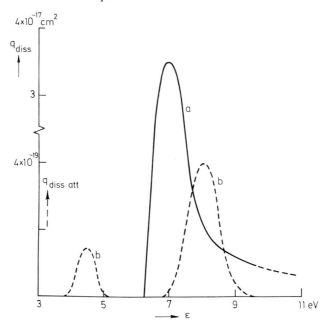

Figure 10.12. Dissociation $q_{diss} = f(\varepsilon)$ (curve a) and dissociative attachment $q_{diss\ att} = f(\varepsilon)$ (curves b) of CO_2.

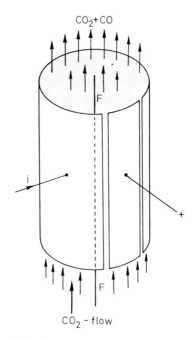

Figure 10.13. Dissociation of CO_2 by electrons of low energy. $F =$ central filament (emitter); split cylindrical Ni anode with C deposited on inner surface.

Instead, the dissociative attachment process received attention

$$e + CO_2 \rightarrow CO + O^- \tag{10.14}$$

Its cross-section as a function of the electron energy is shown in figure 10.12 and was found to rise to $q_{\text{diss att}} \sim 4 \times 10^{-19}$ cm^2. Later measurements of oxygen formed in the positive column of CO_2 and dissociation of CO_2 by electrons at very low pressure emphasized the importance of reaction 10.13. It was shown that the dissociation products of CO_2 resulting from electron collisions of low energy are electrically neutral. This does not contradict the formation of O^- but indicates that the rate of reaction 10.14 is negligible. Figure 10.13 shows an arrangement using an electron-emitting wire at the centre of a split cylindrical aluminium anode internally covered with carbon. The polarity and potential between the wire and the anodes were also varied. By weighing the anode the loss of mass of carbon due to oxidation was determined while a slow axial flow of CO_2 passed through the anode and an electron current crossed radially the gas. The average dissociation cross-section derived was $q_{\text{diss}} \sim 3 \times 10^{-17}$ cm^2 (curve a, figure 10.12) which is about 100 times larger than that of $q_{\text{diss att}}$ (curves b). Hence reaction 10.13 is here the principal process.

CHAPTER 11

welding and processing materials using discharges

11.1. Glow discharge beam welding

We have seen in figure 7.8 (p. 131) that the main component of the discharge current at the cathode of a glow discharge consists of 'beams' of positive ions. If the size of the cathode or the gas pressure for a given current is reduced, the current density and the cathode fall in potential increase. As a result of this, the ion beam energy flowing into the cathode rises and with it its temperature until finally the solid cathode begins to melt and evaporate. Vaporization is distinct from the emission of neutral particles of cathode material resulting from the impact of positive ions ('sputtering'), which sets in at much lower current densities. Such ablation or detachment of neutrals gives rise to a flow of fast atoms or clusters, or a mixture of both. (Sputtering is often used to produce thin films on solid surfaces placed near the cathode and causes deposits on the wall of discharge tubes.)

Two points are important in producing beams of positive ions and electrons: the structure of the beam, and the beam energy and its density. The structure is determined by the geometry of the emitting electrode. The 'beam-focusing', by electric and magnetic lenses, defines the beam's energy density as well as the gas, its pressure and hence the fall spaces. The beam energy depends on discharge voltage, gas pressure, current density and electrode separation. Because of the large mass difference between electrons and ions, and the different collision cross-sections, the design principles applied in ion and electron beam devices can differ substantially. Figure 11.1 shows the $V_d : i$ pressure relation of figure 11.2, a 0·5 kW electron beam system in an air-filled chamber at low pressure.

The method indicated above is used, for example, for welding two metal plates side by side, or two tubes of equal size end to end. The shape of the ion beam was said to depend on the geometry of both anode and cathode, but it is sometimes necessary to apply a third electrode so that the geometry of the emerging ion beam is adjustable by means of controllable voltages applied to the auxiliary electrodes (figure 11.3). By using a low sputtering cathode material (Al)

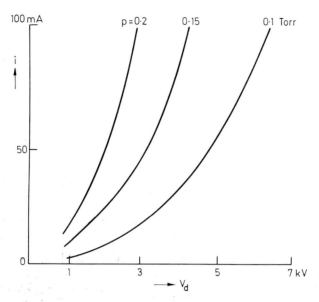

Figure 11.1. Electron beam gun.
Discharge voltage $V_d = f(i)$; i = current in air at different (low) pressures.

this system has the additional advantage of giving freedom in the choice of the nature of the gas.

Figure 11.2 illustrates how a fan-shaped electron beam is formed which heats the separated ends of two metal tubes, D_1 and D_2; when melting starts, the cylinders are forced together (by spring-loaded chunks) whereby a welded joint is obtained. For etching, polishing, machining, sputtering etc., beams of fast positive ions originating in the anode region are preferred, whereas ordinary or post-accelerated electron beams emerging from a curved or hollow cathode of a glow discharge are used for welding. Whether ion or electron beams are extracted, the

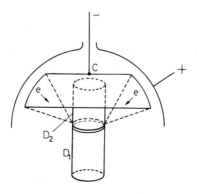

Figure 11.2. Electron beam e, ring-focused, end-on welding of two cylindrical tubes.
+ = anode; C = umbrella-shaped cathode; D_1, D_2 = metal tubes.

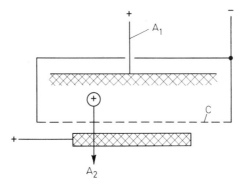

Figure 11.3. Wide positive-ion beam gun.
A_1 and A_2 = main and auxiliary anode; C = wire mesh cathode.

work pieces to be treated or joined have to be placed in a chamber which has to be either partially evacuated or filled with an appropriate gas at low p. Figure 11.2 shows an umbrella-shaped cathode which produces a 'ring-focused' electron beam, e.g., for welding tubes end-on. We see in figures 11.2 and 11.3 that the cathode–anode distances are relatively small. If this arrangement prevents the development of a discharge across the narrow interspace then, in analogy to the fast-rising breakdown voltage $V_s = f(pd)$ to the left of the 'Paschen minimum' (figure 7.2, p. 125), the maintenance voltage for low pd also rises and reaches a value above that of the source voltage. Hence the discharge is forced to take a sufficiently long path between the electrodes (see detour tube, p. 125) to remain stable.

The electron beam in figure 11.2 is focused on the edges of metal tubes which act as an intermediate electrode whose potential is somewhat below that of the anode. To obtain a continuous current in the system, part of the discharge must exist between the work pieces D_1 and D_2 and the anode +. The fast electrons originating from C release secondary electrons at D_1–D_2, in numbers equal to the primaries; this means that the secondary emission coefficient δ (see Chapter 2) is unity and to achieve that, the potential of the insulated metal (work piece) must be approximately 1000–1500 V positive with respect to C, if the discharge is to traverse a vacuum. Because of ionization in the gas, the potential distribution is different. The change can be qualitatively estimated by remembering that the fast positive ions necessary to give the required electron emission from C come partly from the gas contained in the solid angle subtended by the ring focus e at the cathode; the slow positive ions and electrons carry the whole current i to the anode + (see above and Chapter 7). The earthed point (usually chamber and work piece) chosen is entirely a matter of convenience and safety.

The strength of 'beam source' is often expressed in terms of its emission brightness which is given in A cm^{-2} sr^{-1}, the current density per unit solid angle; values are of the order 1–10 and more, depending on what type of beam focusing (electric, space charge or magnetic) is applied. Other points of interest are the power density (in W cm^{-2}) at the work piece (not the cathode) which for electron beams can be up to 10^6 W cm^{-2} for beam voltages of 10–50 kV. Under these circumstances intense X-rays are emitted from solids by the impinging fast electrons; protection against this hazard is demanded by law (shielding).

The efficiency of a beam apparatus, expressed by the ratio of the heat delivered to the work piece and the input power, has been reported to reach over 70 per cent. The physical reason is associated with the fact that, for example, in the case of an ion beam, the potential difference between anode and work piece is larger than that across the diffuse discharge, i.e., between work piece and cathode. Hence the efficiency is over 50 per cent. A similar consideration applies to electron beam arrangement.

At first sight it appears that placing the discharge and the work pieces into a chamber kept at low pressure is a cumbersome requirement. However, the reduction from 1 atm to between 1 and 0·01 Torr for air-filled chambers of reasonable size takes a little time. Rotary oil pumps are usually applied and therefore, in spite of the evacuation, the operation is of practical interest when the superior mechanical properties of the final product are considered. Up to now, continuous steady glow discharges have been assumed to be the beam sources. However, under special circumstances pulsed glow discharges have been successfully used. In this way the input power, which is in general of the order of up to several kW, has been increased to over 1000 kW for pulse lengths of order 1–100 ms.

Arrangements similar to those in figure 11.3 have been used to deposit thin layers of metal, vapour, etc., on the surface of solids (for details see Further Reading).

11.2. Arc discharge welding and cutting

In order to weld metal parts large mechanical forces (cold-welding) or high temperatures must be applied to the joint. Obviously the high temperature of the spots of arcs between metal electrodes are most useful for this purpose. Assume that two flat sheets of metal are to be welded together. In general, the sheets are being made the cathodes of the welding arc while a metal wire is the anode, the tip of which is held a few millimetres above the edges to be welded. Since the spot temperature at the anode $T_a > T_c$, the temperature of the cathode spot, and since the rod material is assumed to be the same as the work piece, the rod, a wire of 1–4 mm diameter and 10–20 cm in length, after striking the arc by contact, is being consumed and thus must be made easily replaceable. This is done by clipping the rod into a holder and tracing the arc by hand along the gap's edges (figure 11.4(a)). Since the anode spot melts the tip of the wire electrode and the cathode spot the edges of the work pieces, the gap between the sheets forms a molten strip. Thus, while the rod electrode is slowly moved along, a narrow short

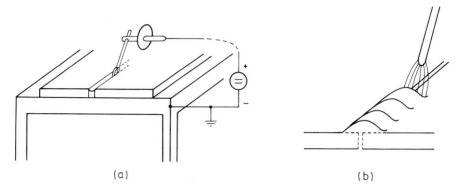

Figure 11.4. D.c. arc welding.
    (a) Hand-welding of 2 plates.
    (b) View of the weld and its surface waves.

liquid pool develops which is continuously filled by droplets shot into it from the tip of the rod. When the inclined rod with its arc spot is moved a small distance the arc length increases slightly. However, the arc spot on the liquid pool appears to be dragged along so that the average arc length, apart from fluctuations, remains the same. Once the arc spot on the melt has left its hottest part, the pool freezes rapidly because cooling by conduction occurs at a high rate. The frozen surface of any weld shows a characteristic pattern (figure 11.4(b)) resembling a set of surface waves on the liquid of the 'capillary type', of wavelength of order 1 mm, controlled by surface tension.

    Such manual arc welding in open air requires currents up to about 300 A; the average arc voltage is between 50 and 80 V with an applied voltage of about 100 V. This simple method is still the most frequently used one in structural and general engineering work, ship building, on railways, and so on. Generators with large armature reaction and transformers with high stray reactance are used as power supplies in order to increase the stability of the welding arcs and thus to avoid re-starting an extinguished arc. Otherwise the mechanical properties of the weld are adversely affected, particularly if the molten slag becomes trapped while the welding pool solidifies. Because of its low average density the slag rises to the pool's surface and, on cooling, forms an insulating layer which prevents or makes it difficult to restrike the arc; the slag has to be chipped away before welding can be continued. To avoid this, instead of bare wire electrodes, chord wires containing a flux in tubular or folded strip electrodes or in a cover applied around the electrodes are used; these may also contain substances which, when heated, give off gases which provide a reducing slag-preventing atmosphere.

    Alternatively, gas-shielded metal arcs, called MIG welding arcs (metal inert gas, figure 11.5) are often used, such as arcs in argon and in helium suitable for non-ferrous, and argon–oxygen mixtures or CO_2 for ferrous metals. From the foregoing it follows that as a result of thermochemical reactions in the weld pool and in the material transferred from the wire to the melt, the composition of the

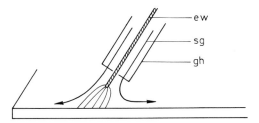

Figure 11.5. MIG (= metal inert gas) welding with a hand-guided arc torch.
ew = electrode wire ≤ 1·5 mm ϕ (automatically fed); sg = shielding gas flow; gh = gun
housing.

weld itself and of the adjoining parts may undergo considerable chemical and
metallurgical changes. For example, the cracking of welds due to hydrogen in
'freezing' steel welds is frequently observed when the hydrogen results from the
water of crystallization contained in the coating of the welding wire. Moreover,
the strength and toughness of welds depends sensitively on traces of de-oxidizing
elements (Si, Ti, Mn, Al) contained in the welding wires. However, except in
automatic welding the property of the weld is patently determined by the skill of
the welder.

In spite of all these complexities it is possible to develop a simple theory
which describes quantitatively the transfer rate of material from a bare wire to a
work piece as a function of the current, given thermal, geometric and arc data.
We assume that the material is conveyed in the form of droplets of given average
size. When the arc spot melts the tip of the wire, a drop forms which increases in size.
The drop's neck, however, shrinks until it is detached from the tip. The elongated
drop closes the arc gap for a short time t_c, but during the subsequent heating,
lasting a time t_h, melting occurs again. If $t_c \ll t_h$, the droplet rate is simply $1/t_h$.

At the start of a heating cycle the stationary arc spot of hemispherical shape
(figure 11.6) gives rise to a temperature field $T(r, t)$ caused by a thermal wave
propagating into the rod. For $r = r_0$, $T = T_0$ which is constant during the interval
t_h. Assuming a constant cathode current density j_c the arc current i determines
r_0. The molten mass of the rod is given by the thickness $(r - r_0)$ of rod material
whose temperature $T_r \geq T_m$, where T_m is the melting point. The partial
differential equation of thermal conduction in radial geometry is

$$\frac{\partial (rT)}{\partial t} = a^2 \frac{\partial^2 (rT)}{\partial r^2} \qquad \text{with} \qquad a^2 = \frac{\lambda}{c_s} \qquad (11.1)$$

where a^2 is the diffusivity in $cm^2\ s^{-1}$, λ the thermal conductivity in $J\ K^{-1}\ cm^{-1}$,
and c_s the specific heat per unit volume. At any r and $t = 0$, $T_r = T_i$, the initial
temperature. The solution of equation 11.1 with $u = (1/2a)(r - r_0)t^{1/2}$ is

$$\frac{r}{r_0} \frac{T_m - T_i}{T_0 - T_i} = 1 - \left(\frac{2}{\pi}\right) \int_0^{u_m} \exp(-u^2)\,du \qquad (11.2)$$

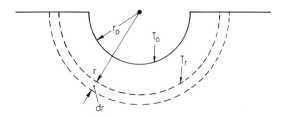

Figure 11.6. To derive the propagation of a hemispherical temperature wave into a semi-infinite solid medium.

where the integral is the error function. Thus, from known values of T_m, T_i, T_0, t_c and r_0 the average mass and volume of a drop can be found. Figure 11.7 shows that the droplet's mass $\bar{G}$ increases with rising droplet rate $(t_h + t_c)^{-1}$ for a given initial temperature T_i. If T_i is assumed to be the average $\bar{T}$ of the spherical shell at $(r_m - r_0)$ at $t = t_h$, it is found that $T_i \sim \frac{1}{2}(3T_m - T_0)$ and its value is 1300–1500 K. $\bar{G}$ has a maximum when $t_h \sim t_c$. Finally, the mass transfer rate from a cathode welding rod of 4 mm diameter as a function of the arc current i (or for $j_c \sim 10^5$ A cm^{-2} or for given r_0) is shown in figure 11.8 and it is seen to vary approximately linearly with i. For an anode rod, because $T_a > T_c$, larger values of $\bar{G}$ (~two-fold) are expected. Both the absolute values of $\bar{G}$ and the polarity effect agree satisfactorily with observations.

Another method, TIG welding (tungsten inert gas), makes use of an arc whose W cathode protrudes from a gun nozzle (figure 11.9). The cathode is very slowly consumed because it is surrounded by a fast-flowing rare gas, usually argon. The work piece is the anode and a 'filler rod' supplies the required material to the welding pool. For TIG welding d.c. or a.c. can be used, because the temperature of the W cathode remains sufficiently high throughout the current

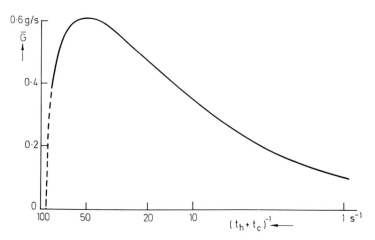

Figure 11.7. Mean droplet mass rate $\bar{G} = f[(t_h + t_c)^{-1}]$ depending on droplet rate. t_h = arc heating interval in seconds; t_c = droplet closure interval in seconds.

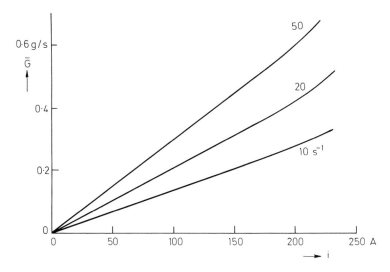

Figure 11.8. Mean droplet mass rate $\bar{G} = f(i)$ for droplet rates of $10–50$ s^{-1}.

period. Also, the arc can be fully pulsed (rate of order 1 s) or partially pulsed, e.g., the current varies between a high and a low value, the latter being so chosen that the weld solidifies between successive pulses.

A variant of the TIG arc is the plasma-welding arc. Here the arc column between the W cathode and the work piece passes through a narrow orifice of a water-cooled nozzle. The arc column is maintained in a fast-flowing rare gas which in turn is surrounded by a flowing shielding gas (figure 11.10). Because of the small orifice the losses in the plasma column and the gas column temperature are relatively high and so are the arc voltage and the energy density at the work piece. At currents above 150 A 'key-hole' welding is possible, i.e., the weld penetrates the whole thickness of the material. The column temperature is $10–20 \times 10^3$ K as compared with $8–15 \times 10^3$ K in the TIG arc. The gas velocities are up to about 10^3 m s^{-1}, i.e., ten times larger, and the power density on the

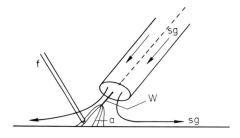

Figure 11.9. TIG (= tungsten inert gas) welding with hand-operated gun.
W = tungsten cathode rod; f = filler rod; a = arc; sg = shielding gas.

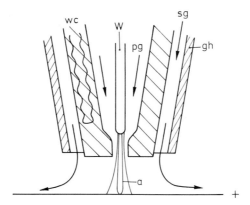

Figure 11.10. Plasma welding with tungsten.
pg = plasma gas; wc = water cooling; a = arc; gh = gun housing; sg = shielding gas.

work piece is up to ten times that of a TIG arc. The arc voltage of a 250 A plasma arc is about 40 V, but for the TIG arc it is only up to 20 V.

Arc cutting requires the removal of material in the molten state. To cut a thick metal sheet efficiently, the width of the cut, i.e., the volume of material lost, has to be kept small and the cutting speed high. The high-power plasma arc, because of its thin gas column and large power density, seems to be most suitable.

11.3. Laser welding and cutting

Infra-red laser beams of high power, particularly those produced by glow discharges in gas mixtures of CO_2-N_2-He (in a ratio of approximately $1:2:3$) at total gas pressure of 10–100 Torr, are of interest for processing materials. They are not only used for cutting, drilling, welding and annealing metals, but also ceramics, papers and textiles. Laser beams of up to about 10 kW continuous power, and many orders of magnitude higher power from pulsed beams, are now available. However, the beam intensity must exceed approximately 10^5 W cm^{-2} of absorbed power to balance melting, evaporation, radiation, reflection losses, ablation (droplets) and ionization. The present upper limit of continuous intensity is about 10^9 W cm^{-2}. Single pulse and high-frequency pulse operation is of advantage when the temperature rise has to be confined to a narrow region in the work piece. This is necessary in order to reduce stresses in the material, to avoid chemical changes and phase transformations. $1-10^2$ kJ cm^{-3} per pulse or μs pulses of 10^2-10^3 kJ are within our compass.

Consider now the forces which occur when a narrow laser beam of diameter less than 0·1 mm falls perpendicularly on a metal sheet of several millimetres thickness. Ordinary incoherent light of moderate intensity of wavelength 10 μm suffers negligible absorption (≤ 5 per cent) by ordinary untreated metal surfaces. The temperature of the metal surface will therefore only rise moderately. The absorption can be increased by blackening or roughing the surfaces or by oxidation. However, a coherent light source is always associated with a beam of high intensity, and so even a few per cent of the power converted into heat causes surface temperatures to rise at a rate of the order of 10^4 K s^{-1} for 10 kW cm^{-2} absorbed power density which extends to a depth of a few tenths of a millimetre. Thus less than 0·1 s after applying the beam, melting and evaporating sets in when a beam of spot size 1 mm is applied, and a little later a hole forms. When the hole depth becomes about equal to the spot diameter, evaporation and ablation no longer occur into a solid angle 2π but are confined to a smaller solid angle because the solid walls of the hole act like a gun barrel. Because of the high beam intensity the initially large reflection of the beam decreases, while the density of the vapour and the amount of molten globules and clamps expelled rises. This somewhat heterogeneous atmosphere begins later on to block the hole and the beam intensity deep in the hole is attenuated.

At the same time absorption of quanta result in excitation and ionization of the hot vapour cloud above the hole and still later a nearly fully ionized plasma cloud of metal vapour in air is formed. The whole process takes place so rapidly that for intensities of about 10^7 W cm^{-2} a laser–ionization wave of supersonic speed propagates along and in a direction opposite to that of the beam. When a steady state is reached the power delivered by the beam to the plasma is balanced by plasma energy losses due to radiation, thermal conductivity, convection and so on, but the distribution of the losses depends on the power density. From this account it follows that the upper limit of the beam intensity which produces the ignition of a plasma in air corresponds to the breakdown field of an electric discharge in air free of impurities. It has been found that air of one atmosphere requires an intensity of 10^9 W cm^{-2}. The associated electric field in the laser beam E is about 10^6 V cm^{-1}, a value near that needed to induce field ionization in air. However, when the irradiated solid supplies metal vapour at sufficient rate, the starting power level p of the air-metal vapour plasma is considerably lower, about 10^5 W cm^{-2}, than that of pure air, which seems to agree with observations. Since the transfer of power from the beam to the target via the vapour plasma is a wave problem, its solution describes the formation and maintenance of the beam plasma. This is of considerable importance in assessing the coupling (matching) of the beam power to the initially solid target.

The beam emerging from a CO_2 laser discharge in a low-pressure gas, which contains a relatively large amount of He and smaller amounts of N_2 and CO_2, is produced as follows. The flowing laser gas mixture is led through one or several extended d.c. glow discharges, whereby the direction of current is either parallel or perpendicular to that of the flow. The light from the optical cavity or cavities, i.e., the luminous discharge volume bounded by special mirrors, after being carefully aligned and adjusted, is led through a window separating the low-pressure chamber from the atmospheric air to a focusing mirror which produces a small spot on the target. In certain applications the spot is moved over the target by a movable mirror.

For what reason, besides power conversion efficiency, has the $10 \cdot 6 \,\mu m$ radiation of CO_2 been selected as the carrier of high-power laser beams? This triatomic molecule of linear symmetric structure (figure 11.11) is, in the electronic ground state, an emitter of a multitude of lines and bands, and electronic excitation of CO_2 would make its emission spectra still more complex. The answer is that N_2 was the first molecular gas shown to emit a near infra-red laser beam when subjected to pulsed excitation, which causes a transition up to the vibrational level 1 followed by a transition to level 0 by molecular collisions. The filling of the level $v = 1$ occurs by electron collisions, either directly from $v = 0$ or via higher intermediate states of N_2, or by cascading from higher electronic states N_2^*. All vibrational levels have a relatively long natural life because of their metastability, provided that collisions are absent or rare. Thus, with hindsight, it is not surprising that in a steady-state system—not in a system in thermodynamic equilibrium—the population density N_1 in a higher vibrational state can be made larger than the concentration N_0 in the lower state. This population inversion means that instead of $N_1/N_0 \propto \exp\left[-(\varepsilon_1 - \varepsilon_0)/kT\right]$ we may write either $N_1/N_0 \propto \left[-(\varepsilon_1 - \varepsilon_0)/(-kT)\right]$ and use the negative temperature concept (which is not in fashion any longer) or simply evaluate N_1/N_0 from the rate equations. Of course an inversion can also be obtained by using a high draining (de-excitation) rate for the lower state instead of a high feeding rate to a higher state only. However, there are several more restrictions imposed to obtain laser action. Besides a large population in the upper state, which requires a high electron concentration, there are the resonance condition of the optical cavity, the extracted beam power and the losses. An upper state can also be populated by 'optical pumping', i.e., absorption of radiation (photo-excitation) by particles in a lower level; in the CO_2 laser radiationless energy transfer by $N_{2\,vib}$ colliding with CO_2 occurs.

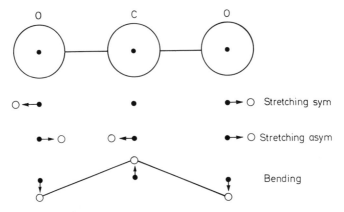

Figure 11.11. Stretching and bending modes of CO_2 in the electronic ground state of the molecule.

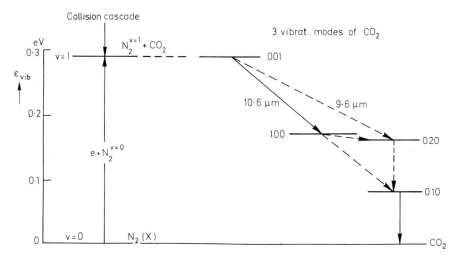

Figure 11.12. Transfer of the vibrational excitation $v = 1$ of N_2 to the vibrational levels of CO_2.

The strength of an emitted laser beam depends on the transition probability of the pair of levels involved. In CO_2 (figure 11.12) the vibrational level of

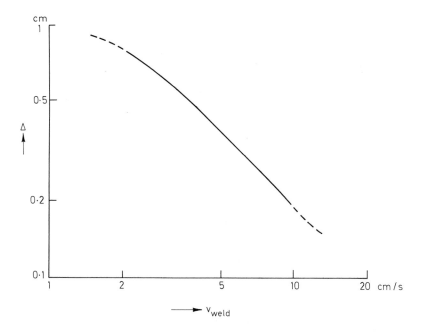

Figure 11.13. Laser weld penetration depth $\Delta = f(v_{weld})$. v is the welding speed in steel for a 7 kW continuous wave CO_2 laser. At low v_{weld}, Δ saturates because the width of the hole produced increases the depth.

asymmetric stretching 001 has nearly (to $\Delta\varepsilon \sim 1$ meV) the same potential energy ε as N_2, $v = 1$, namely 0.29 eV (energy-resonance, large transfer cross-section). Thus the numerous $N_2^{v=1}$ molecules readily transfer by collisions (with a probability of about 10^{-3}) energy to CO_2 molecules which vibrate briefly in the 001 mode until, induced by interaction, they radiate coherently at $\lambda = 10.6\,\mu m$ and continue vibrating in the symmetrical stretching mode 100 at 0.17 eV, a level that is drained by transition to lower levels which vibrate by bending. We have now arrived at energy levels of less than 0.1 eV and, since room temperature corresponds to about 0.03 eV, we note that with a laser gas at several hundred kelvin above room temperature these lowest levels may be partly filled by collisions with the gas. The transition $001 \rightarrow 020$, $\lambda = 9.6\,\mu m$, has to be kept weak so as not to de-activate the 001 population. Figure 11.13 shows the drop in penetration depth in steel at 7 kW when the welding speed is raised.

Finally, the role of He has to be discussed briefly. Its large thermal conductivity facilitates gas cooling. As an atomic constituent of the gas, it raises the electron temperature and the fraction of fast electrons; the light He atoms depopulate the lower levels of CO_2, thermalize rotational levels and increase the number of fast electrons by shaping the distribution curve.

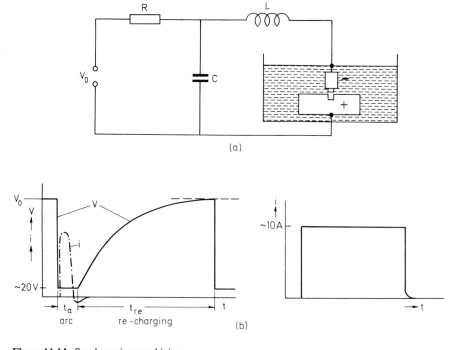

Figure 11.14. Spark erosion machining.
  (a) Circuit and tank for pulsed arc and work piece.
Figure 11.14. (b) Constant energy and current pulse; where anode drop energy $E_a \propto V_m$, molten volume; $E_a \propto h^3$, erosion depth h.

11.4. Spark erosion machining

Electric arcs of short duration have been used recently for machining, i.e., drilling (ordinary, tapered or profiled holes), for producing narrow grooves, profiled cavities, and so on in standard as well as in 'difficult' materials.

The principle of the spark erosion technique, developed first by Lasarenko (1938), is as follows: a d.c. arc about 1 mm long is started by making contact between an electrode (a cathode of W–Cu, graphite or cast iron) and the anode workpiece. The arc is maintained for 2–$2000\,\mu s$ and then interrupted and restarted; the pulse frequencies are 3×10^2–$3 \times 10^5\ s^{-1}$. Owing to the explosive nature of the pulsed arc, a small amount of molten material is ejected from the crater in the workpiece (figure 11.14) in the form of droplets. The arc gap is submerged in a paraffin-filled tank and the particles swim in the liquid. If necessary the eroded material can be removed by a flow of the liquid. The arc is supplied either with constant current pulses of 3–20 A, often from a d.c. transistorized voltage source of up to about 100 V, or by constant energy pulses of up to $0{\cdot}5$ J and down to 10^{-5} J for finishing, whereby a charged condenser acts as the source. For finishing work high frequencies and low power are obviously more suitable, whereas large removal rates of material demand low frequency, high power and sometimes negative polarity applied to the workpiece. Figure 11.14 gives the circuits and some data. The electrode erosion method is a slow thermal process in which large mechanical forces are absent.

CHAPTER 12

atmospheric and cosmic plasmas

12.1. General remarks

There exist a variety of ionization phenomena which belong to a group called cosmic plasmas. The first question is whether ionized gas, such as air of very low density far above the surface of the Earth, is truly representative of a plasma? One answer is that in order to qualify for 'plasmaship', the Debye length Λ (in m) $\simeq 70(T_e/N_e)^{1/2}$, with the electron temperature T_e in K and the concentration N_e in m^{-3}, must be larger than a characteristic linear plasma dimension, such as the radial thickness of the region considered. Since we know that N_e and T_e are sufficiently uniform over distances of several kilometres, whereas the largest values of the Debye length Λ are of the order of 10 m, the conditions for treating the ionized regions as plasmas are fulfilled. A cube of edge length $\Lambda = 10$ m, though containing only 10^7 ion pairs m^{-3}, holds altogether 10^{10} ion pairs. However, it cannot be assumed that in astrophysics we shall always encounter ideal or fully ionized plasmas.

Instead of describing plasmas by electron concentration N_e and electron temperature T_e, two other plasma parameters are used. The electron plasma frequency (in Hz) $f_e \simeq 9(N_e)^{1/2}$, with N_e in m^{-3}, is of the order $10^8–10^{10}$ s^{-1} in laboratory plasmas, where N_e can be, say, 10^5 m^{-3} in outer space and f_e only some 10 kHz. On the other hand, weak magnetic fields reveal their presence in the electron cyclotron frequency $f_c \simeq 1\cdot8 \times 10^7 B$ with B in gauss (see Chapter 4). In outer space, where the magnetic field B has dropped to values of less than 10^{-5} gauss (compared to about $0\cdot4$ gauss at the Earth's surface), f_c is of the order 100 Hz. Hence at large B, f_c may approach f_e.

Other critical factors are the time constants associated with changes in local plasma conditions. These have to be clearly distinguished from plasma changes which are caused by variations in distant sources such as the Sun (sunspots). In this case the travel time of a disturbance to reach the 'absorption' region, i.e., the local plasma, may play a significant role.

In a sense cosmic plasmas present simpler systems than extraterrestrial or laboratory plasmas because of the absence of electrodes, surfaces or walls. There, in general, electric charges can become neutralized at a large rate and excited particles deactivated; also, mechanical and chemical changes in the solid can lead to evaporation, ablation or other kinds of 'sputtering' that may influence the character of the plasma. It is for this reason that the effect of cosmic dust in cosmic plasmas has attracted much attention; yet too little is known quantitatively about it to indicate its relative importance.

12.2. The Sun and the solar plasma

Since the Sun is the most effective plasma source of the terrestrial atmosphere it may be useful to illustrate first the Sun's atmosphere and give some numerical data which should help in comparing cosmic and laboratory plasmas.

Figure 12.1 shows a section of the Sun's atmosphere. The surface (black body) temperature is about 5800 K; it is maintained by radioactive processes within the body of the Sun. The most probable group of processes is the 'proton–proton chain'. This reaction, in which protons, $_1^1 H$, form $_2^3 He$ followed by $_2^4 He$ and finally protons again, reads

(a) $\qquad _1^1 H + _1^1 H \rightarrow (_2^2 He) \rightarrow _1^2 H + e^+$

(b) $\qquad _1^2 H + _1^1 H \rightarrow _2^3 He$

(c) $\qquad _2^3 He + _2^3 He \rightarrow (_4^6 Be) \rightarrow _2^4 He + _1^1 H + _1^1 H$

$(_2^2 He)$ and $(_4^6 Be)$ are temporary compound nuclei, e^+ is a positron, $_1^2 H$ is deuterium and $_2^3 He$ a rare isotope of ordinary He. As to the fate of e^+, we note that since the reactions (a) to (c) take place in the interior of the Sun where matter is fully ionized, there is a huge population of free electrons e^- present and thus positron–electron annihilation occurs

$$e^+ + e^- \rightarrow 2\gamma$$

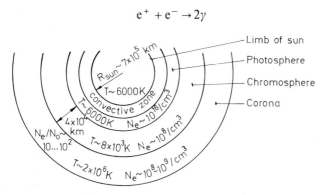

Figure 12.1. Regions surrounding the sun.

where 2 γ-ray quanta of $0 \cdot 5$ MeV each moving in opposite directions, are emitted and finally absorbed in the surroundings.

As figure 12.1 shows, the Sun is surrounded by a convective zone whose properties are somewhat uncertain, followed by the photosphere, the chromosphere and the corona in which the plasma temperature is about 2×10^6 K. This is contrary to what one expects, since temperatures are usually found to decrease with increasing distance from a hot body. The temperature rise from the Sun's surface to the corona by two to three orders of magnitude is still a puzzling phenomenon. It has been suggested that overheating of the coronal gas is due to shock waves generated in the photosphere.

Figure 12.1 also gives some information about the electron concentration and temperature in the various zones, and their radial extent in units of R_{Sun}. It has to be remembered that the Earth is about $200R_{\text{Sun}}$ from the Sun and is, so to speak, immersed in the coronal or solar plasma which produces effects that will be discussed now.

12.3. The weak plasma of the ionosphere (the Kennelly–Heaviside or E layer, the D layer and the Appleton or F layers)

Qualitative considerations

It has already been mentioned in Chapter 1 that the upper atmosphere surrounding the Earth is partly ionized and that some of the molecules therein are in excited states. Also, certain molecules, consisting of up to three atoms, such as ozone and carbon compounds, are found preferentially in the upper layers, i.e., in regions of lower gas density, at higher concentrations than on the Earth's surface. It has been known for a long time by mariners that occasionally in far northern and southern latitudes luminous effects—the aurorae—appear in the sky high above the Earth; at the same time the strength and direction of the magnetic field of the Earth changes in an unpredictable manner, interrupting wire and wireless telegraphy, particularly during periods of strong visible disturbances at the surface of the Sun. All the evidence collected points towards the presence of free electric charges of both signs in the upper atmosphere. These were subsequently located at heights of about 100 km and more above the Earth. Later, more detailed investigations disclosed that there existed not one, but several ionized layers in the upper atmosphere, one above the other. A similar argument on the absorption of the Sun's radiation suggested the existence of a layer of ozone (O_3) which results from photochemical changes (photodissociation of O_2 by ultraviolet rays); the ozone layer was found to be at around 30 km, i.e., between the ionized layers and the Earth. The height distribution of atmospheric data is given in tables 12.1 and

Table 12.1. Distribution of atmospheric parameters with height h.

h (km)	p (Torr)	T (K)	N (m^{-3})	λ (cm)	ρ (kg m^{-3})
0	760	288	$2 \cdot 6 \times 10^{25}$	$7 \cdot 4 \times 10^{-6}$	$1 \cdot 2 \times 10^{1}$
10	200	223	$8 \cdot 6 \times 10^{24}$	$2 \cdot 2 \times 10^{-5}$	$4 \cdot 1 \times 10^{-1}$
20	43	214	$1 \cdot 9 \times 10^{24}$	$1 \cdot 0 \times 10^{-4}$	$8 \cdot 9 \times 10^{-2}$
50	0·75	276	$2 \cdot 4 \times 10^{22}$	$8 \cdot 5 \times 10^{-3}$	$1 \cdot 2 \times 10^{-3}$
100	5×10^{-4}	230	$1 \cdot 8 \times 10^{19}$	$9 \cdot 0$	$8 \cdot 8 \times 10^{-7}$
200	4×10^{-7}	700	$5 \cdot 0 \times 10^{15}$	3×10^{4}	$1 \cdot 6 \times 10^{-10}$
500	2×10^{-9}	1000	$1 \cdot 0 \times 10^{13}$	$\sim 10^{6}$	$2 \cdot 0 \times 10^{-13}$

12.2 (N, ρ and λ are number density, mass density and mean free path, respectively).

From what has been explained in the previous chapters, it can be deduced that atmospheric electrification requires the presence of a gas of sufficiently large density. On that score ionization should become the more intense the nearer it is to the Earth's surface, where the density of the atmosphere is highest. On the other hand, whatever the nature of the agency which produces ions and electrons, if the agents originate from far regions of the universe and propagate towards the Earth, it can be assumed that absorption will reduce their strength along the path, being more powerful farther away from Earth. Therefore, since the intensity increases with the distance from the Earth's surface, whereas the gas density decreases with height at a certain height, the degree of ionization must pass a maximum.

In fact this had been thought to be established, quite unexpectedly, during the early experiments around 1900 which Marconi conducted in this country. A radio transmitter of several hundred metres in wavelength at Poldhu in Cornwall emitted signals which were recorded well below the optical horizon some 4000 km away, at Signal Hill in Newfoundland. According to the theory accepted at that time, the signal strength, being a decreasing function of the distance, should only be detectable at points a fraction of that distance from the transmitter. The explanation of this effect was that the aerial of the transmitter emitted signals hemispherically in all directions, except vertically upwards. The electromagnetic waves travelling obliquely skywards were deflected (refracted) by an ionized layer, say 100 km or more above the Earth, and from there travelled down to far places without being unduly absorbed. To a first approximation, such ionized layers act as reflectors for radiation, provided that the wavelength is sufficiently short. From

Table 12.2. Relative gas composition of the atmosphere by volume (%).

h (km)	O_2	C	N_2	N
90	21	0	78	0 (1% Ar)
100	20	1	79	0
150	4	26	65	5
200	2	46	41	11
400	0	69	3	27

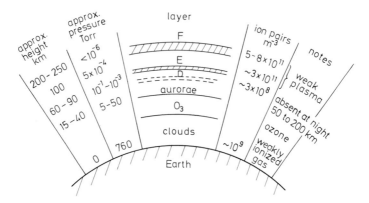

Figure 12.2. The Earth and its surrounding regions.

later measurements it transpired that several such 'reflecting' layers exist and that their strength varies with the time of the day. The agency maintaining them is the Sun. The parts of the Sun's spectrum which cause ionization in the upper atmosphere are the far-ultraviolet and the soft X-ray region. In terms of wavelengths, this would mean the spectral regions below about 150 nm, also called the vacuum-ultraviolet, and the adjoining soft X-ray region at about 10 nm or less.

Some properties of the various layers, their height and names are illustrated in figure 12.2.

Quantitative considerations

An approximate treatment of the steady state of a single ionized layer (*E* region) is as follows. Assume that ionizing radiation, such as soft X-rays which are emitted by the Sun, penetrates the upper regions of the Earth's atmosphere because the gas density in the outer space is so low that, in spite of the large distances, the intensity of radiation is still high. If the gas temperature *T* in space were constant (in fact it decreases somewhat with rising height) then gas density $\rho \propto p$, the gas pressure. The dependence of *p* on the height *h* above the ground (figure 12.3) follows from the well-known argument that the change in pressure d*p* in an element d*h* of unit area is

$$dp = -gp\, dh \tag{12.1}$$

where *g* is the gravitational acceleration.

Using the ideal gas law, *m* being the mass of a molecule, *N* the concentration and $\rho = Nm$ the mass density

$$p = NkT = (\rho/m)kT \tag{12.2}$$

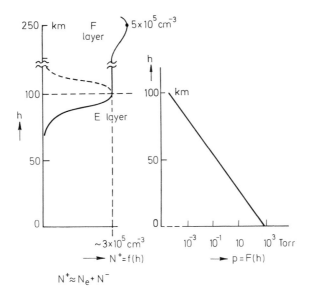

Figure 12.3. Positive ion concentration N^+ and gas pressure $p = f(h)$. h = height above Earth, 0 = sea level.

Substituting p in equation 12.1 and integrating between $h = 0$ where $p = p_0$ ($\rho = \rho_0$) and h, one obtains, for constant temperature

$$p_h/p_0 = \rho_h/\rho_0 = \exp{-(mgh/kT)} = \exp{-(h/H)} \qquad (12.3)$$

i.e., the barometric law, showing that the density depends on the ratio of potential to kinetic energy, or on the 'scale height' $H = kT/mg$. At about 100 km, where T is about 200 K, the 'equivalent' gas pressure p_h is about 10^{-4} Torr and H is about 6 km.

If I_h is the intensity of ionizing radiation at h in quanta $cm^{-2} s^{-1}$, i.e., the flux density, and assuming continuous absorption (which is in general not the case) expressed by a coefficient $\mu = a\rho_h$ (in cm^{-1}) where a is constant, then the change in intensity dI_h in dh is

$$dI_h = \mu I_h \, dh \qquad (12.4)$$

With $h \to \infty$ and $I_h \to I_\infty$, we find from equations 12.3 and 12.4

$$I_h/I_\infty = \exp{[a\rho_0 H \exp{(-h/H)}]} \qquad (12.5)$$

Equation 12.5 is a monotonically rising double-exponential function which for $h = 0$ has the constant value $\exp{(-a\rho_0 H)}$ and which for large h approaches 1 asymptotically.

If absorption of radiation would lead to ionization only, the ionization rate $dN_i/dt = dI_h/dh$, N_i being the ion concentration (in cm^{-3}). However, only a fraction b of absorbed radiation ionizes where $b\mu = q_i N_h$, q_i being the photo-ionization cross-section (in cm^2) of the radiation of quantum energy $h\nu$ and N_h, the concentration (cm^{-3}) of the molecules. Hence with radiation of normal incidence

$$\frac{dN_i}{dt} = q_i N_h \frac{dI_h}{dh} \tag{12.6}$$

From equations 12.4, 12.6, 12.2 and 12.3, we find for the ionization rate at the relative height $z = h/H$

$$\frac{dN_i}{dt} = [q_i a I_\infty (\rho_0^2/m)] \exp - [(a\rho_0 H) \exp(-z) + 2z] \tag{12.7}$$

To evaluate equation 12.7 numerically, we find for $H = 6 \times 10^5$ cm and $\bar{T} = 200$ K the relative height z_{max} at which the rate is a maximum by differentiating the exponent and equating it to zero. This gives for $a\rho_0 = 500$ $z_{max} = \ln(a\rho_0 H/2) = 18$ and thus the maximum rate dN_i/dt lies theoretically at a height h_{max} of about 108 km, which is near the earlier determined value of 100 km. For $T\lambda_{Sun} = 6 \times 10^3 \times 10^{-5} = 6 \times 10^{-2}$ K cm, the flux density of quanta (between 0 and 100 nm) I_∞ is about 3×10^{14} cm^{-2}. Since the first square bracket in equation 12.7 $q_i N_0 a\rho_0 I_\infty$ is about $10^{-17} \times (2 \times 10^{-19}) \times (5 \times 10^2) \times (3 \times 10^{14}) = 3 \times 10^{19}$, the ionization rate at z_{max} is

$$\frac{dN_i}{dt} \sim 3 \times 10^{19} \exp(-40) \sim 60 \text{ ion pairs cm}^{-3} \text{ s}^{-1}$$

If dissociative recombination ρ_{diss} were the principal loss, its rate

$$\frac{dN_i}{dt} = \rho_{diss} N_e N^+ = \rho_{diss} N_e^2 \tag{12.8}$$

and we obtain in the steady state at z_{max} with ρ_{diss} of about 10^{-8} cm^{-3} s^{-1} a number density of $(60/10^{-8})^{1/2} \sim 8 \times 10^4$ ion pairs cm^{-3} which compares well with the observed density of about 10^5 electrons cm^{-3}.

There are other aspects which are of considerable interest, such as the time required to set up the E layer (τ of order 0.5 h) and the width of the E layer, which was thought to be of the order 20 km, measured between two points both at one half of the maximum concentration of charge, and its variation with time. But one of the important questions concerns the presence of more than one ionized layer, as seen in figure 12.2: the D layer below and the F layer above the E layer. The general pattern which causes an ionized layer to develop appeared to be the increase in gas density with decreasing height, associated with an increase in absorption of that part of radiation which produces ionization. This, however, would not explain why more or less well separated layers seem to develop, were it not for strong absorption bands in the gas mixture at or around that particular height when ionizing radiation, pre-filtered by the gas traversed, is present in sufficient quantity. Or one might ask whether the nature of the 'radiation' which gives rise to the various layers is different.

Superficially the whole picture seemed to be consistent. Even the numerical treatment, as indicated above, looked quite reassuring, until in the 1960s some cracks appeared in the edifice. These were caused by 'minor details' such as the width of the layers absorbed, their patchiness and horizontal movements, and most importantly, the absence of any direct experimental evidence of the existence of electron density maxima. Only the rising part of the E layer curve leading to a supposed maximum had been actually observed.

With the advent of satellites and space vehicles the situation changed completely because, instead of transmitting and receiving signals from the ground, the space vehicle could carry a transmitter, and in this way a maximum in electron density could be easily spotted. The result was surprising. Briefly, no maxima around 100 km have been observed except one high up at about 250 km, in the F layer. Instead, the new physical picture is roughly that shown in figure 12.3. The so-called ionized layers are recognizable at the lower side by kinks in charge density; moreover, where a maximum was thought to lie, a transition into the next higher region sets in, exemplified by the steep increase in rate of rise in electron density.

Let us now summarize the present view, though details are by no means well established.

The former D layer

This is the lowest ionized region whose activity reveals itself in reflecting high-frequency signals. The electron number density is about 10^3 cm^{-3}, the corresponding critical frequency is about 10^5 s^{-1}, the height 60–90 km. This region disappears at night because the agency producing ions in daylight is absent. Although ionizing X-rays of 0·1–1 nm may be present, the main agency is thought to be the H_α (Lyman-α) line at 121·6 nm (see Chapter 3) which is emitted by excited H atoms in the region of the Sun, and which penetrates the upper ionosphere until it is absorbed in the lower and denser part of the atmosphere. This is because the absorption of H_α in N_2 is very weak, and O_2 is strongly dissociated at these heights, while the remaining O_2 has a gap (window) in the absorption spectrum at 121·6 nm. Since this wavelength corresponds to 10·2 eV and the ionization potential V_i of O_2 is 12·1 V, no direct photo-ionization can occur, nor can N_2 (with $V_i = 15·6$ V) be photo-ionized by absorption of H_α radiation. However, it appears that sufficient NO is formed, for example by the process

$$N + O \rightarrow NO$$

$$NO + e \rightarrow NO^+ + 2e$$

With nitric oxide ($V_i = 9·3$ V) present, the ionization rate is probably large enough to maintain a charge density of up to 10^3 electrons cm^{-3} against the recombination loss by $e + NO^+ \rightarrow N + O$ which is of the dissociative type. The contribution to the loss by ion–ion recombination is quantitatively not yet known because of lack of reliable data on O_2^- and O^-.

The former E layer

Between 90 and 150 km extends an ionized region which was once thought to have parabolic electron density distribution $N_e(h)$ with height but is now found to consist of a rising part followed by a plateau; above about 150 km N_e rises to still higher densities which form the F region. In the E region the electron density is up to about 2×10^5 electrons cm^{-3} during daytime and falls by a factor of about 100 during the night. Ultraviolet radiation between 90 and 102·6 nm (the latter being the Lyman-α line) and soft X-rays between 1 and 17 nm are thought to be the photo-ionizing sources producing O_2^+ and NO^+ ions which are lost by dissociative electron–ion recombination. It has to be noted that N_2 requires radiation of $\lambda \leq 80$ nm to be ionized and its intensity below 150 km is too low to add noticeably to the total rate of ionization. The reflective (refractive) property of this region, which is found to be split into two, corresponds to a critical frequency of a few megahertz.

The F layers

A final step appears in the curve electron density versus height at a value of about 4×10^5 electron cm^{-3} and a critical frequency of about 9 MHz. From there the density rises first slowly (F_1 region) and later faster until it reaches a maximum at about 5×10^5 cm^{-3} (F_2 region) lying at 250 km; all the values quoted refer to average daytime conditions. By this we mean that periods during which the frequency of Sun spots and thus of excessive ionization is high are excluded. From the maximum to about 1000 km the electron density decreases almost linearly with height. Within the F regions the ionization is likely to be due to ultraviolet radiation of 17–90 nm. The principal ion is O^+ but at heights above 500 km O^+ and H^+ balance the electron charge. Also, the electron density (and therefore the total ion density) is found to decrease at night by a factor of approximately five, showing that agencies other than the Sun's radiation are more powerful than the ultraviolet. Electrons of energy less than 10^3 eV have been suggested to be responsible for the strong background ionization observed during darkness.

From the account given it transpires that the ionosphere is a feeble gaseous plasma which is maintained by an Earth-directed flux of electromagnetic radiation (or light quanta) and of fast charged particles which arrive from outer space, one source being the Sun. The ionosphere cannot be described by a series of parallel ionized layers of definite thickness but rather by a series of regions of varying charge density (but with zero net space charge), depending on the height above ground, which vary with time, i.e., with the position of the Sun relative to the Earth. However, the observed data also depend on the Sun's activity (sunspots); associated with it is a large increase in X-ray emission, and a solar 'wind' which carries the plasma, mainly horizontally, around the Earth. The non-uniformity of the incoming flux produces irregularities in ionization rates, patches of larger charge densities and plasma clouds. The conclusion must be drawn that—like the weather at the Earth's surface—the 'ionospheric weather' is subject to continuous changes, sometimes accompanied by violent disturbances such as magnetic storms, and by spectacular auroral displays, lighting up the sky at heights of 80–130 km with the peak in the E region. The green O line at 557·7 nm and the blue $(N^+)^*$ line at 391·4 nm, listed in tables of spectra as OI and NII, are prominent at lower, and the red lines at 630 and 636·4 nm at higher levels. The excitation is probably due to electrons of $10–10^4$ eV energy.

Measurements with space vehicles indicate the distribution and motion of the geomagnetically trapped charges which survive periods of up to several months in space. Their paths are confined to regions surrounding the Earth which are known as the van Allen radiation belts. The fast component of charged particles consists of electrons of up to $10^5–10^6$ eV and positive ions of several 10^8 eV; the former circulate at heights of a few Earth radii above ground, i.e., 20 000–30 000 km in the Sun's direction, while the latter, mainly protons, are moving in the 'inner radiation belt' at a height of about one half of the Earth's radius (≤ 5000 km). The slow component consists of charges which yield the larger part of the ionization, move mainly in closed orbits and are finally captured by hitting the Earth. The changes in the outer belt are due to cosmic rays coming from outer space, and ionization produced by them is found to be large near the magnetic poles and lower near the equator.

Outside the radiation belts there is a flow of solar plasma of low charge density, the solar wind, which originates in the corona (figure 12.1). Its field of flow is distorted by the geomagnetic field as seen in figure 12.4, which in turn changes into an unsymmetrical dipole field with the Earth immersed in an egg-shaped magnetic cavity, the magnetosphere; on the side facing the sun the geomagnetic field appears to be compressed. The solar plasma thus flows around

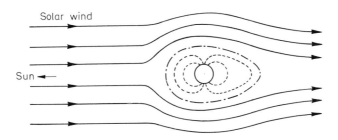

Figure 12.4. Solar wind and the Earth.
————— = flow lines; – · – · – = magnetosphere; – – – – – = geomagnetic lines.

the Earth as an air flow that encounters a spherical obstacle. Flow speeds of 500 km s^{-1} have been recorded and proton densities of about 3 ions cm^{-3}.

Large-scale experiments in space, such as 'Argus', give further confirmation of the long life of the charged particles in the radiation belts. Several small atomic bombs were detonated at large heights. The released relativistic electrons resulting from β decay of the fission products such as ^{92}Kr were trapped by the geomagnetic field and oscillated along the magnetic lines between its two mirror points (see Chapter 4). In addition, the electrons were found drifting eastwards within a thin shell around the Earth. The life of the charges, the extent of the shell and the accompanying auroral luminescence were measured by satellites and rockets confirming, for example, the expected shell thickness and the stability of the 'motionless' shell.

12.4. Lightning discharges and thunderstorms

General assessments

Over 200 years ago Benjamin Franklin in the USA and d'Alibard in France carried out the first serious experiments on the electrification of clouds. In 1752 Franklin flew a kite into a thundercloud and observed sparks between the kite's wire and the ground (having the good fortune to escape unhurt). Since then balloons, aircraft and radar have been used to investigate the production, location and distribution of electric charges in clouds as well as the regeneration time of a charge after a lightning discharge has passed between two clouds or between a cloud and the ground. One result, confirmed by recent observations, was that in air the mean electric field before a lightning stroke seldom exceeds 3–4 kV cm^{-1}, i.e., 10 per cent of the breakdown field on the ground. The potential differences between cloud and ground would amount to, say, 10^9 V. Since the cloud ceiling is about 10 km, at which height gas density is about one third of that at ground level, breakdown there would require a field of about 10 kV cm^{-1} and a higher value for the mean field would ensue. Whatever the assumptions made (within reason) the available field is three times smaller than the expected one, and the question why remains unanswered. Actually, the discrepancy is larger than the above figures suggest, because the cloud ceiling is often much lower than 10 km. Another point concerns the maximum stored charges released in a single stroke, which is of the order 10–30 C.

Here we pause for reflection to consider the situation before a thundercloud develops. Since the terrestrial atmosphere contains a net ion space charge which, near ground level, is of the order 700 (positive) ions cm^{-3} we have, with the (vertical) current density j about 10^{-6} A cm^{-2} and the ion mobility about 1 cm s^{-1} (V cm^{-1})$^{-1}$, an electric field E of about 1 V cm^{-1} to maintain j. The direction of E is towards the ground, since the air has a larger number of positive ions (singly charged) than negative. This is the 'fine-weather field' which drives the negative ion upwards. Very different in topology and magnitude are the measured fields inside and at the boundary of thunderclouds.

It is often glibly stated that this charge passes from the cloud to the ground. If we think in terms of closed circuits, an explanation is required of how this charge can circulate through 'the circuit'. In nature the cloud–ground system (figure 12.5(*a*)) is a charged capacitance with one electrode near the ground, but the path of the current is obscure. Figure 12.5(*b*) shows that the current *i* associated with the lightning circulates through the discharge, up in one direction and down through the space surrounding the discharge, as displacement current (dashed), in the opposite direction. This will be the case whether the discharge current is oscillatory or highly damped and singly pulsed. The current in the main stroke can be of the order 10^4–10^5 A and the magnetic fields near it about 100 G. The duration is of the order 10^{-4} s. All the figures quoted are, of course, subject to considerable statistical uncertainties; hence data often vary in order of magnitude. The time interval between two successive strokes, the 'regeneration time' of charge, has been observed to be of the order 20 s and more; but in giant storms, cloud electrification can be restored in 2–3 s. The vertical length of ordinary lightning is usually 2–4 km and slightly more for cloud-to-cloud discharges.

Production, separation, collection and neutralization of charges

It is useful to introduce first the concept of cells in thunderclouds. These are cylindrical air spaces of considerable height, the active centres of thunderstorms that have been observed by aircraft and radar. In these cells, about 1 km in diameter and 10 km in height, air currents flow initially vertically upwards with speeds of perhaps 30 m s^{-1} for, say, about 15 minutes preceding a storm. When later rain or hail falls to the ground, these warm up-draughts melt the ice and are cooled by rain drops of increasing size and by evaporation. The fall in

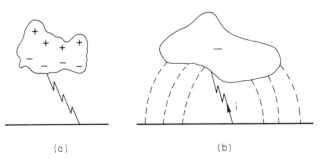

(a)   (b)

Figure 12.5. (*a*) Thunderstorm cloud and lower charge distribution.
(*b*) Lightning and current distribution. $- - - - =$ current return paths.

temperature reduces the up-draught, which cannot support drops and hail any longer. At the cell's top the temperature can be as low as 253–263 K. Finally, a strong down-draught develops which accelerates the flow of droplets.

The generation of electric charges in these cells is still a mystery. The simplest charge distribution found in a cloud is shown in figure 12.5(a). It is seen that the negative charge is, in general, nearer to the ground. However, sometimes there is an additional smaller pocket of positive charges at the base of the cloud. Several points should be emphasized.

(a) The negative charge in the cloud is much larger than the positive charge at the top of the cloud. If this were not so, the electric field between a cloud and the Earth might be too small to produce the high voltage which starts a lightning to ground.

(b) After the discharge, which uses only a fraction (say one-half) of the total negative charge, regeneration of the cloud charge may take only a short time. It is known that discharges between the central negative charge and the upper positive charge take place. If the field value above and below the negative charge were the same, the starting voltage of an upward discharge in stagnant gas would be lower because of the lower pressure, although the lower temperature at the top of the cloud makes the difference in the field per unit gas density (E/N) somewhat smaller. The net result would be a preference of lightning in the cloud. However, the distribution of charges with height and the corresponding 'internal fields' in the cloud may be strongly influenced by the vertical flows of air, the presence of ice pellets and other cloud parameters which finally determine the path of the lightning.

(c) The point or area from which a lightning travels to the ground is determined by the charge accumulation at the lower boundary of the cloud, which is thought to produce an intense field of strong non-uniformity, such as is known for solid electrodes of conical shape. One problem is to find the field of charges which are in the process of being separated (say vertically) by aerodynamic flow currents, by solving the corresponding space charge equation that applies to a system similar to the van de Graaff machine (p. 168) with the upper electrode replaced by a charged cloud. Another problem is the collection of charges of equal sign into the lightning channel. The solution will show what fraction of total charge can be used in a single stroke, since it is not possible to collect quickly all charges from large distances and force them into a narrow discharge path.

We shall not discuss here the finer details dealing with the development of the lightning discharge. These aspects are treated in some of the more recent accounts cited in Further Reading of this Chapter and Chapter 8.

FURTHER READING

Chapters 1 and 2

Aston, F. W., 1942, *Mass Spectra and Isotopes* (London: Arnold).

Badareu, E. and Popescu, I., 1965, *Gaz Ionisés* (Paris: Dunod) (in French).

Brown, S. C. 1967, *Basic Data in Plasma Physics*, 2nd edition (Cambridge, Mass.: MIT Press).

Callear, H. B., Devenport, C. P. and Kendall, D. R., 1981, *Journal of Chemical Physics* **61**, 65 (associative excitation of Hg).

Cobine, J. D., 1943, *Gaseous Conductors*, 3rd edition (New York: McGraw–Hill).

Clampitt, R. and Jefferies, D. K., 1978, *Nuclear Instruments and Methods* **149**, 734 (liquid metal ion emitter).

de Groot, W. and Penning, F. M., 1933, *Handbuch der Physik* **23**, 1 (in German).

Delcroix, J. L., 1966, *Physique des Plasmas*, Volumes 1 and 2 (Paris: Dunod) (in French).

Druyvesteyn, M. J. and Penning, F. M., 1940, *Review of Modern Physics* **7**, 230.

Duckworth, H. E., 1958, *Mass Spectroscopy* (Cambridge: Cambridge University Press).

Emeleus, K. G., 1951, *Conduction of Electricity through Gases* (London: Methuen).

Faraday, M., 1844, *Researches in Electricity* (London: Dent).

Frank-Kamenetskii, D. A., 1972, *Plasma* (New York: Plenum).

Granowski, W. L., 1955, *Elektrischer Strom im Gas* (Berlin: Akademischer Verlag) (in German).

Hasted, J. B., 1972, *Physics of Atomic Collisions*, 2nd edition (London: Butterworth).

Kapzow, N. A., 1955, *Elektrische Vorgänge in Gasen* (Berlin: Verlag der Wissenschaften) (in German).

Laporte, M., 1948, *Decharge Electrique dans le Gas* (Paris: Colin) (in French).

Llewellyn Jones, F., 1955, *Ionization Avalanches* (London: Methuen).

Loeb, L. B., 1939, *Fundamental Processes* (New York: Wiley; Berkeley: University of California Press, 1955).

Meek, J. M. and Craggs, J. D. (eds.), 1978, *Electrical Breakdown* (New York: Wiley).

Mitchener, M. and Kruger, C. M., 1973, *Partially Ionized Gases* (New York: Wiley).

Mott, N. F., 1972, *Elementary Quantum Mechanics*, Wykeham Science Series No. 22 (London: Taylor & Francis).

Prewett, P. D. and Jefferies, D. K., 1980, "Low energy ion beams", in *Bath Conference*, No. 54, edited by I. H. Wilson and K. G. Stephens (Bristol, UK: Institute of Physics).

Seeliger, R., 1934, *Physik der Gasentladungen* (Leipzig: Barth) (in German).

Tan, K. L. and von Engel, A., 1968, *Journal of Physics*, *D* **1**, 258 (associative ionization cross-section of Hg).

Thomson, J. J. and Thomson, J. P., 1933, *Conduction of Electricity through Gases*, Volumes 1 and 2 (Cambridge: Cambridge University Press).
Townsend, J. S., 1915, *Electricity in Gases* (Oxford: Clarendon Press).
von Engel, A., 1965, *Ionized Gases*, 2nd edition (Oxford: Oxford University Press).

Chapter 3

Bates, D. R. and Bederson, B., 1979, *Advances in Atomic Physics*, Volumes 14 and 15 (New York: Academic Press).
Capitelli, M. and Dilonardo, M., 1978, *Journal of Chemical Physics* **30**, 95 (O_2).
Capitelli, M. and Dilonardo, M., 1978, *Journal of Chemical Physics* **34**, 193 (H_2).
Capitelli, M. and Dilonardo, M., 1978, *Revue de Physique Appliquée* **13**, 115 (N_2).
Chatjian, A. and Cartwright, D. C., 1981, *Physical Review* **A23**, 2178 (q_e^* for Ar).
Cottrell, A. H., 1964, *Mechanical Properties of Matter* (New York: Wiley).
de Heer, F. J., 1975, *International Journal of Radiation Physics and Chemistry* **7**, 137 (Super-excited molecular state).
Dekker, A. J., 1964, *Solid State Physics* (London: Macmillan).
Dolder, K. T. and Peart, B., 1976, *Reports on Progress in Physics* **39**, 697.
Dunn, G. H., 1980, *JILA Report* No. 2201 (Boulder, Col.: Joint Institute for Laboratory Astrophysics) (Electron-ion collisions).
Flowers, B. H. and Mendoza, E., 1970, *Properties of Matter* (New York: Wiley).
Gilardini, A. L., 1973, *Low Energy Electron Collisions* (New York: Wiley).
Gill, R. D. (ed.), 1981, *Plasma Physics and Fusion Research* (Summer Schools 1978–80; Culham, UK) (New York: Academic Press).
Hasted, J. B., 1979, *Advances in Atomic and Molecular Physics* **15**, 205 (Double charge transfer).
Herzberg, G., 1944, *Atomic Spectra* (New York: Dover).
Johnson, R. C., 1950, *Atomic Spectra*, 2nd edition (London: Methuen).
Kistemaker, J., 1981, *Natuurkunde* **90**, 71 (Lecture on atomic collision physics 1960–81).
Kittel, C., 1966, *Introduction to Solid State Physics*, 3rd edition (New York: Wiley).
Kuhn, H. G., 1969, *Atomic Spectra*, 2nd edition (London: Longman).
Ladd, M. F. C. and Lee, W. H., 1969, *Modern Physical Chemistry* (Harmondsworth, UK: Penguin).
Latham, R. V., 1981, *High Voltage Vacuum Insulation* (New York: Academic Press).
Lawson, J. D., 1977, *Physics of Charged Particle Beams* (Oxford: Oxford University Press).
McDaniel, E. W., 1964, *Collision Phenomena in Ionized Gases* (New York: Wiley).
Massey, H. S. W., 1976, *Negative Ions*, 3rd edition (Cambridge: Cambridge University Press).
Massey, H. S. W., 1979, *Atomic and Molecular Collisions* (London: Taylor & Francis).
Massey, H. S. W. and Berhop, E. H. S., 1969–74, *Electronic and Ionic Impact Phenomena*, Volumes 1–5 (Oxford: Oxford University Press).
Massmann, P. *et al.*, 1981, in *10th Conference on Controlled Fusion, Moscow*, Volume 1 p. H–20 ($D^+ \rightarrow D^-$).
Mott, N. F., 1930, *Outline of Wave Mechanics* (Cambridge: Cambridge University Press).
Phillips, J. A. and Tuck, J. L., 1956, *Review of Scientific Instruments* **27**, 97 (H source).
Present, R. D., 1958, *Kinetic Theory of Gases* (New York: McGraw–Hill).
Rapp, D., Englander-Golden, P. and Briglia, D. D., 1965, *Journal of Chemical Physics* **42**, 4081.

Riviere, R. C., 1969, *Solid State Surface Science* **1**, 179 (Work functions).
Sakuntala, M. and von Engel, A., 1960, *Journal of Electronics* **9**, 31 (Charge transfer and dissociative excitation).
Schulz, P., 1968, *Elektronische Vorgänge in Gasen* (Karlsruhe, FRG: Braun) (in German).
Smirnov, B. M., 1982, *Negative Ions* (New York: McGraw–Hill).
Smithells, C. J., 1976, *Metals*, 5th edition (London: Butterworth) (Work functions).
Tabor, D., 1979, *Gases, Liquids and Solids*, 2nd edition (Cambridge: Cambridge University Press).
Tan, K. L. and von Engel, A., 1971, *Proceedings of the Royal Society of London* **A324**, 183 (Super-elastic electron collisions).
Thorne, A. P., 1972, *Spectrophysics* (London: Chapman & Hall).
Vidaud, P. H. and von Engel, A., 1978, *Journal of Physics*, *D* **11**, 1597 (Electron back-diffusion).
von Ardenne, M., 1956, *Tabellen*, Volume 2 (Berlin: Verlag der Wissenschaften), p. 636 (in German).

Chapter 4

Bull, C. S., 1966, *Fluctuations of Electric Currents* (London: Butterworth).
Chapman, S. and Cowling, T. G., 1970, *Non-uniform Gases*, 3rd edition (Cambridge: Cambridge University Press).
Cosslett, V. E., 1950, *Introduction to Electronic Optics*, 2nd edition (Oxford: Oxford University Press).
Dutton, J., 1975, *Journal of Physical and Chemical Reference Data* **4**, 577 (Electron swarms).
Hayishi, M., 1981, *Transport Cross-sections* (Nagoya, Japan: Institute of Plasma Physics).
Jost, W., 1952, *Diffusion in Solids, Liquids, Gases* (New York: Academic Press).
Klemperer, O., 1972, *Electron Physics*, 2nd edition (London: Butterworth).
Klemperer, O. and Barnett, M. E., 1971, *Electron Optics*, 3rd edition (Cambridge: Cambridge University Press).
Krupenie, P. H., 1972, *Journal of Physical and Chemical Reference Data* **1**, 423 (Spectrum of O_2).
Little, P. F. and von Engel, A., 1954, *Proceedings of the Royal Society of London* **A224**, 209 (Hollow cathode discharge; refined evaluation: P. F. Little, D. Phil. Thesis Oxford University 1952).
Navinsek, B. (ed.), 1976, *Proceedings of the 8th International Summer School, University of Ljubljana* (Ljubljana, Yugoslavia: Stefan Institute).
Ollendorf, F., 1932, *Potentialfelder der Elektrotechnik* (Berlin: Springer) (in German).
Smith, D., Dean, A. G. and Adams, N. G., 1972, *Zeitschrift für Physik* **253**, 191 ("diffusion cooling" $[T_e < T_{gas}]$ occurs in pure rare gas after-glows after say a few ms when T_e reaches saturation value; cause: at low (pD_a) energetic electrons ambipolarly diffusing to the wall penetrate the sheath and deliver neutralization energy. The electron gas is thus cooled transiently).
Smythe, W. R., 1968, *Static and Dynamic Electricity*, 3rd edition (New York: McGraw–Hill).
von Engel, A., 1965, *Ionized gases*, 2nd edition (Oxford: Oxford University Press).

Chapter 5

Alfvén, H. and Fälthammar, C. G., 1963, *Cosmical Electrodynamics*, 2nd edition (Oxford: Oxford University Press).

Bekefi, G., 1966, *Radiation Processes in Plasmas* (New York: Wiley).

Clemmow, P. C. and Dougherty, J. P., 1969, *Electrodynamics of Particles and Plasmas* (Reading, Mass.: Addison–Wesley).

Franklin, R. N., 1976, *Plasma Phenomena in Gas Discharges* (Oxford: Oxford University Press), p. 124 (Electron wave damping).

Franklin, R. N., Hamberger, S. M., Lampis, G. and Smith, G. J., 1975, *Proceedings of the Royal Society of London* **A347**, 1.

Hamberger, S. M., 1973, *Contemporary Physics* **14**, 559.

Sheffield, J., 1975, *Plasma Scattering of Electromagnetic Radiation* (New York: Academic Press).

Shohet, J. L., 1971, *The Plasma State* (New York: Academic Press), p. 230 (Landau damping).

Spitzer, L., 1962, *Physics of Fully Ionized Gases*, 2nd edition (New York: Interscience).

Chapter 6

Alfvén, H. and Fälthammar, C. G., 1963, *Cosmical Electrodynamics*, 2nd edition (Oxford: Oxford University Press).

Alpher, R. A. and White, D. R., 1959, *Physics of Fluids* **2**, 162.

Burhorn, F., 1959, *Zeitschrift der Physik* **115**, 42.

Cook, M. A. and McEwan, W. S., 1958, *Journal of Applied Physics* **29**, 1612.

Dalgarno, A. and Lewis, J. T., 1957, *Proceedings of the Royal Society of London* **A240**, 285.

Forrest, J. R. and Franklin, R. N., 1968, *Proceedings of the Royal Society of London* **A305**, 251 (Diamagnetic column).

Francis, G., 1960, *Ionization Phenomena in Gases* (London: Butterworth).

Franklin, R. N., 1976, *Plasma Phenomena in Gas Discharges* (Oxford: Oxford University Press), p. 47 (Diamagnetic plasma).

Klein, S., 1974, *Compte Rendu Hebdomadaire des Séances de l'Académie des Sciences, B* **277**, 541 (in French; Image transmission through water by ultrasonic waves).

Knechtli, R. C. and Wada, J. Y., 1961, *Physical Review Letters* **6**, 215.

Langmuir, I., 1929, *Physical Review* **33**, 954.

Lin, S. C., Resler, E. L. and Kantrowitz, A., 1955, *Journal of Applied Physics* **26**, 95.

Moisan, M., Zakrzewzki, Z. and Pantel, R., 1979, *Journal of Physics, D* **12**, 219 (Surface wave produced plasmas).

Rompe, R. and Steenbeck, M., 1939, *Ergebnisse der Naturwissenschaften* **18**, 257 (in German).

Sakuntula, M., von Engel, A., and Fowler, R. G., 1960, *Physical Review* **112**, 1.

Spitzer, L. 1962, *Physics of Fully Ionised Gases*, 2nd edition (New York: Interscience).

Steenbeck, M., 1939, *Wissenschaftliche Veröffentlichungen aus den Siemens-Werken*, **18/3**, 94 (in German; Dalton's Law).

Troppmann, M., 1970, *Zeitschrift für Naturforschung* **25a**, 504 | (in German; alkali plasma diode).

von Engel, A., 1965, *Ionized Gases*, 2nd edition (Oxford: Oxford University Press).

von Engel, A. and Steenbeck, M., 1932, 1934, *Elektrische Gasentladungen*, Volumes 1 and 2 (Berlin: Springer) (in German).

Wallis, G., 1982, *Beiträge aus der Plasmaphysik* **22**, 295 (in English; Soft X-ray emission from pulsed laser-generated plasmas).

Chapter 7
(see also Further Reading, Chapters 1 and 2)

Badareu, E. and Ciobotaru, D., 1963, in *Proceedings of the 7th International Conference on Ionized Phenomena in Gases*, Volume 1, edited by P. Hubert (Saclay, France: Centre d'Etude Nucléaires), p. 97 (Hollow cathode discharge Xe); see also 1964, *International Journal of Electronics* **17**, 529).

Brown, S. C., 1966, *Introduction to Electrical Discharges* (New York: Wiley–Interscience).

Butler, H. S. and Kino, G. S., 1963, *Physics of Fluids* **6**, 1346 (Plasma sheath in r.f. fields).

Carré, B., Roussel, F., Breger, P. and Spiess, G., 1981, *Journal of Physics, B* **14**, 4271 and 4289 (Approaching a pulsed fully ionized gas).

Cobine, J. D., 1941, *Gaseous Conductors*, 2nd edition (New York: McGraw–Hill).

Emeleus, K. G., 1981, *Journal of Physics, D* **14**, 2179 (Cathode regions glow discharge).

Emeleus, K. G., 1982, *International Journal of Electronics* **52**, 407 (Anode glows).

Francis, G., 1956, *Handbuch der Physik* **22**, 53.

Francis, G., 1960, *Ionization Phenomena in Gases* (London: Butterworth).

Gill, P. E. and Webb, C. E., 1977, *Journal of Physics, D* **10**, 299 (Negative glow and electron energy distribution).

Goldan, P., 1970, *Physics of Fluids* **13**, 1055 (Electric field near a 'floating' metal plate in a plasma).

Haque, R. and von Engel, A., 1974, *International Journal of Electronics* **36**, 239 (N_2^* metastable concentration by chemical colourimetry).

Hirsh, M. N. and Oskam, H. J. (eds.), 1978, *Gaseous Electronics*, Volume 1 (New York: Academic Press).

Hoyaux, M. F., 1968, *Arc Physics* (Berlin: Springer).

Lafferty, J. M. (ed.), 1980, *Vacuum Arcs* (New York: Wiley).

Llewellyn Jones, F., 1966, *Glow Discharge* (London: Methuen).

Lochte-Holtgreven, W. (ed.), 1968, *Plasma Diagnostics* (Amsterdam: North Holland).

Lombardi, G. G., Smith, P. L. and Parkinson, W. H., 1979, *Journal of the Optical Society of America* **69**, 1289.

MacDonald, A. D., 1966, *Microwave Breakdown in Gases* (New York: Wiley).

Nasser, E., 1971, *Fundamentals in Gaseous Ionization* (New York: Wiley).

Parker, A. B. and Johnson, P. C., 1975, *Proceedings of the Royal Society of London* **A325**, 511 (V_s at low pressure).

Raether, H., 1964, *Electron avalanches* (London: Butterworth).

Robson, A. E. and von Engel, A., 1957, *Proceedings of the Royal Society of London* **A242**, 214 (Excitation theory of vapour arcs).

Smith, D. and Plumb, J. C., 1972, *Journal of Physics, D* **5**, 1226 (Probes in passive plasmas).

Somerville, J. M., 1959, *Electric Arc* (London: Methuen).

Swift, J. D. and Schwar, M. J. R., 1970, *Electrical Probes* (London: Iliffe).

von Hippel, A. R., 1965, *Molecular Designing* (Cambridge, Mass.: MIT Press), p. 167 (V_s at high pressure).

Zhu, S. L. and von Engel, A., 1981, *Journal of Physics, D* **14**, 2225 (Atmospheric arcs of vanishing length).

Chapter 8

(see also Further Reading, Chapter 12)

Allen, T. K. and Tait, J. H., 1975, in *Proceedings of the International Conference on Uranium Isotope Separation* (London: British Nuclear Engineering Society), paper 6.

Blyth, A. R. (ed.), 1975, *Proceedings of the Conference on Static Electrification* (London: Institute of Physics), p. 182.

Breare, J. M., 1973, in *Cosmic Rays at Ground Level*, edited by A. W. Wolfendale (London: Institute of Physics), ch. 10.

Chalmers, J. A., 1967, *Atmospheric Chemistry* (London: Pergamon).

Chen, I. and Mort, J., 1975, *Physics of Xerographic Receptors* (New York: Academic Press).

Chubb, J. N., 1973, *Culham Laboratory Report*, No. P338 (Ignition hazards during tank washing).

Chubb, J. N. *et al.*, 1973, *Nature* **245**, 206.

Deutsch, W., 1933, *Physikalische Zeitschrift* **34**, 448 (in German).

Deutsch, W., 1933, *Annalen der Physik* **16**, 588 (in German).

Farquhar, R. L., 1977, Ph.D. Thesis (Bath University) (Flame radiation detector).

Illingworth, A. J. and Krehbeil, J. H., 1981, *Physics in Technology* **12**, 122.

Jackson, R. N. and Johnson, K. E., 1974, *Advances in Electronics* **35**, 191.

Moore, A. D. (ed.), 1973, *Electrostatics and its Applications* (New York: Interscience).

Musgrove, P. J. and Wilson, A. D., 1972, *Energy Conversion* **12**, 21.

Musgrove, P. J. and Wilson, A. D., 1973, *Electronics and Power* **19**, 327.

Pankow, J. I. (ed.), 1980, *Display Devices* (Berlin: Springer).

Rose, H. E. and Wood, A. J., 1966, *Electrostatic Precipitation* (New York: Dover).

Schaffert, R. M., 1975, *Electrophotography*, 2nd edition (London: Focal Press).

Schofield, J. M. S., 1975, in *12th Conference on Phenomena in Ionized Gases*, Part 1, edited by J. G. A. Hölscher and D. C. Schram (Amsterdam: North Holland), p. 73.

Schofield, J. M. S., 1976, in '4th Conference on Gas Discharges', *I.E.E. Publication* No. 143 (London: I.E.E.), p. 397.

Sharpe, J., 1955, *Nuclear Radiation Detectors* (London: Methuen).

Smith, J., 1975, *IEEE Transactions* **ED22**, 642 (Display panels).

Stubbs, R. J. and Breare, J. M., 1969, in *11th Conference on Cosmic Rays, Budapest* (Budapest: Academie Kiado) (Sealed spark chambers).

Taylor, D., 1951, *Measurement of Radioisotopes* (London: Methuen).

Uman, M. A., 1969, *Lightning* (New York: McGraw–Hill).

Weston, G. F., 1980, *Physics in Technology* **11**, 218 (Displays).

White H. J., 1963, *Industrial Electrostatic Precipitation* (New York: Addison–Wesley).

Wynakker, M. M. B. and Granneman, E. H. A., 1980, *Zeitschrift für Naturforschung* **35a**, 883 (in English; Mass separation in a weakly ionized plasma centrifuge).

Chapter 9

Arzimovich, L. A., 1965, *Elementary Plasma Physics* (New York: Blaisdell).

Bickerton, R. J., 1976, *Essays in Physics*, Volume 6 (New York: Academic Press).

Bickerton, R. J. and Keen, B. E., 1974, in *Plasma Physics Conference Series*, Volume 20, edited by B. E. Keen (London: Institute of Physics), p. 243.

Conlon, T. W., 1982, *Contemporary Physics* **23**, 353 (Surfaces doped with radioactive atoms).

Cook, I., 1975, *Encyclopaedic Dictionary of Physics*, Supplement to Volume 5 (Oxford: Pergamon), p. 341.
Francis, G., 1967, *Science Progress* **55**, 53.
Motz, H., 1979, *Laser Fusion* (New York: Academic Press).
Pease, R. S., 1975, in *12th International Conference on Phenomena in Ionized Gases*, Part 2, edited by J. G. A. Hölscher and D. C. Schram (Amsterdam: North Holland).
Robson, A. E., 1962, *Encyclopaedic Dictionary of Physics*, Volume 7 (Oxford: Pergamon), p. 326.
Semat, H., 1964, *Atomic and Nuclear Physics*, 4th edition (London: Chapman & Hall), p. 580.
Spalding, I., 1974, *Laser Interactions*, Volume 3 (New York: Plenum), p. 775.
Teller, E. (ed.), 1981, *Fusion* (New York: Academic Press).
Tokamak design and JET-Tokamak, *Culham Laboratory Annual Reports*, 1979–1981 (London: H.M.S.O.).

Chapter 10

Aiba, T. and Freeman, M. P., 1974, *I.E. Review* **13**, 179.
Baddour, R. F. and Timmins, R. S. (eds.), 1967, *Application of Plasmas to Chemical Processing* (Oxford: Pergamon).
Barton, M. J. and von Engel, A., 1970, *Physics Letters* **A32**, 173 (CO_2 dissociation).
Boschke, F. L. (ed.), 1980, *Plasma Chemistry*, Volumes I and II, Topics in Current Chemistry Nos. 89 and 90 (Berlin: Springer).
Capitelli, M. and Dilonardo, M., 1978, *Journal of Chemical Physics* **34**, 193.
Capitelli, M. and Dilonardo, M., 1978, *Revue de Physique Appliquée* **13**, 115.
Cernogora, G., Hochard, L., Touzeau, M. and Matos Ferreira, C., 1981, *Journal of Physics*, *B* **14**, 2977 (A-metastable state population of N_2).
Claxton, K. T., 1961, *Journal of the Imperial College Chemical Engineering Society* **14**, 58 (CO_2 dissociation).
Claxton, K. T., 1964, *Carbon* **1**, 495.
Corrigan, S. J. B. and von Engel, A., 1958, *Proceedings of the Royal Society of London* **245**, 335 (H_2 dissociation).
Deegan, P. J. and Emeleus, K. G., 1948, *Annalen der Physik* **3**, 149 (O_3).
Fauchais, P. and Rakowitz, J., 1979, in "Proceedings of 14th International Conference on Phenomena in Ionized Gases", *Journal de Physique* **40**, Supplement Volume 2, C7–289.
Glockler, G. and Lind, S. C., 1939, *Electrochemistry of Gases* (New York: Wiley).
Gross, B., Grycz, B. and Miklossy, K., 1969, *Plasma Technology* (London: Iliffe).
Haque, M. R., 1974, *International Journal of Electronics* **36**, 239 (N_2 metastables).
Hirth, M. 1981, *Beiträge aus der Plasmaphysik* **21**, 1 (in German; O_2 production in silent discharges).
Hollahan, J. R. and Bell, A. T., 1974, *Techniques and Applications of Plasma Chemistry* (New York: Wiley).
Kaufman, F., 1969, *Advances in Chemistry Series* **80**, 29.
Kelly, A. J., 1976, *Journal of Applied Physics* **47**, 5264 (liquid metal spraying).
Keiffer, L. J., 1969, *JILA Report* No. 5264, 6 January 1969 (Boulder, Col.: Joint Institute for Laboratory Physics).
Kirkby, P. J., 1911, *Proceedings of the Royal Society of London* **85**, 151 (H_2O discharge synthesis).

Korvin, K. K. and Corrigan, S. J. B., 1969, *Journal of Chemical Physics* **50**, 2570 (CO_2).

Laidler, K. J., 1955, *Chemical Kinetics of Excited States* (Oxford: Oxford University Press) (P. E. surfaces of CO_2).

Lawton, J. and Weinberg, F. J., 1969, *Electrical Aspects of Combustion* (Oxford: Oxford University Press).

Lunt, R. W., 1940, *Journal of the Royal College of Science* **10**, 123.

McTaggart, F. K., 1967, *Plasma Chemistry in Electrical Discharges* (Amsterdam: Elsevier).

Massey, H. S. W., 1960, *Electronic Impact Phenomena*, Volumes 2 and 5 (Oxford: Oxford University Press).

Niehaus, A., 1967, *Zeitschrift für Naturforschung* **22a**, 690.

Polak, L., 1971, *Reactions under Plasma Conditions*, Volume 2 (New York: Wiley), ch. 12.

Polak, L., 1974, *Pure and Applied Chemistry* **39**, 307.

Polak, L., 1977, in *13th International Conference on Phenomena in Ionized Gases, Berlin* (Berlin, GDR: Akademie der Wissenschaften), p. 251.

Reed, T. B., 1967, *Advances in High Temperature Chemistry* **1**, 260.

Sabadil, H., Bachmann, P. and Kastelewicz, H., 1980, *Plasmaphysik* **20**, 283 (in German; O_3 reaction kinetics in O_2 flow).

Spence, D., 1981, *Journal of Chemical Physics* **74**, 3898 (new O_2 P.E. curves).

Suhr, H., 1973, in *11th International Conference on Phenomena in Ionized Gases, Prague*, Volume 2, edited by L. Bekarek and L. Laska (Prague: Czechoslovak Academy of Science, Institute of Physics), p. 413.

Tachibana, K. and Phelps, A. V., 1981, *Journal of Chemical Physics* **75**, 3315.

Warburg, E., 1927, *Handbuch der Physik* **14**, 167 (in German).

Wright, A. N. and Winkler, C. A., 1968, *Active Nitrogen* (New York: Academic Press).

Yaron, M. and von Engel, A., 1975, *Chemical Physics Letters* **33**, 316 (O_3 dissociation).

Yaron, M. and von Engel, A., 1977, in *3ème Symposium International de Chemie de Plasmas, Limoges*, Volume 1 (Limoges: University Dept. of Chemistry), p. G.2.24.

Zajonc, A. G. and Phelps, A. V., 1981, *Physical Review* **A23**, 2479 (Non-radiative transport of Na* excitation).

Chapter 11

Agarwal, P. K., Bark, E. L. and Weigand, A. J., 1982, *A.I.A.A. Journal* **20**, 1137 (Ion beam surface texturing).

American Welding Society, 1978, *Welding Handbook*, Volumes 1 and 2, 7th edition, (London: Macmillan).

Bickes, R. W. and O'Hague, J. B., 1982, *Review of Scientific Instruments* **53**, 585 (Duo-picatron ion source).

Davidson, A., 1974, *Physical and Chemical Fabrication Techniques*, Handbook of Precision Engineering, Volume 4 (London: Macmillan).

Dugdale, R. A., 1971, *Glow Discharge Material Processing* (London: Mills & Boon).

Dugdale, R. A., 1974, *Journal of Materials Science* **10**, 896 (Glow discharge welding).

Fink, J. H., 1975, *Welding Journal* **54**, Supplement, 137.

Fowler, M. C. and Smith, D. C., 1975, *Journal of Applied Physics* **46**, 138.

Herriott, D. R., 1982, *Journal of Vacuum Science and Technology* **20**, 781 (Electron beam lithography).

Houldcroft, P. T., 1967, *Welding Processes* (Cambridge: Cambridge University Press).

Houldcroft, P. T., 1975, *Welding Processes* (Oxford: Oxford University Press).

Kaufman, H. R. and Robinson, R. S., 1982, *A.I.A.A. Journal* **20**, 745 (Broad-beam industrial ion sources).
Knight, C. J., 1982, *A.I.A.A. Journal* **20**, 950 (Transient vapourization, surface into vacuum).
Megaw, J. H. P. C., 1981, *Culham Report*, No. CLM-P649 (Steel plates with unmachined edges laser welded).
Megaw, J. H. P. C. and Spalding, I. J., 1976, *Physics in Technology* **7**, 187.
Metcalfe, J. C. and Quigley, M. B. C., 1975, *Welding Journal* **54**, March.
Nighan, W. L. and Wiegand, W. J., 1974, *Applied Physics Letters* **25**, 633.
Nighan, W. L. and Wiegand, W. J., 1975, *Applied Physics Letters* **26**, 554.
Pease, R. S., 1959, in *Plasma Physics (Varenna Summer School)* (Bologna: Zanchelli), p. 158.
Pike, E. R. (ed.), 1976, *High Power Gas Lasers (Capri Summer School)* (London: Institute of Physics).
Rykalin, N. N. and Nikolaev, A. V., 1970, in *International Institute of Welding Colloquium, Lausanne* (Lausanne, Switzerland: International Institute of Welding).
Say, M. G., 1973, *Electrical Engineers Reference Book*, 13th edition (London: Butterworth).
Solotych, B. N., 1955, *Elektrofunkenbearbeitung* (Berlin, GDR: VEB Verlag Technik).
Szekely, J. and Poveromo, J. J., 1974, *Metallurgical Transactions* **A5**, 289.
von Engel, A., 1937, *Wissenschaftliche Veröffentlichungen aus den Siemens-Werken* **16**, 70 (Arc-welding theory).
Willett, C. S., 1974, *Gas Lasers: Population Inversion Mechanisms with Emphasis on Selective Excitation Processes* (Oxford: Pergamon).
Yagi, S. and Janaka, M., 1979, *Journal of Physics, D* **12**, 1509 (Mechanism of O_3 generation in air).

Chapter 12

Alfvén, H. and Arrhenius, G., 1976, *Evolution of the Solar System* (Washington, D.C.: N.A.S.A.).
Boley, F. I., 1966, *Plasmas: Laboratory and Cosmic* (New York: Van Nostrand Reinhold).
Boyd, R. L. F., 1975, *Space Physics: Study of Plasmas in Space* (Oxford: Oxford University Press).
Golde, R. H. (ed.), 1977, *Lightning* (London: Academic Press).
Malan, D. J., 1964, *Physics of Lightning* (London: English Universities Press).
Mason, B. J., 1972, *Proceedings of the Royal Society of London* **A327**, 433.
Massey, H. S. W. and Bates, D. R. (eds.), 1982, *Atmospheric Physics and Chemistry* (New York: Academic Press).
Meek, J. M. and Graggs, J. D. (eds.), 1978, *Electrical Breakdown of Gases* p. 432.
Schonland, B. F. J., 1956, *Handbuch der Physik* **22**, 576.

Examples, exercises and essays

Ferrari, R. L. and Jonscher, A. K. (eds.), 1973, *Problems in Physical Electronics* (London: Pion), ch. 5, p. 181.
Nasser, E., *Fundamentals of Gaseous Ionization* (New York: Wiley).
von Engel, A., 1965, *Ionized Gases*, 2nd edition (Oxford: Oxford University Press).

INDEX

239